T0138003

Studies in Big Data

Volume 101

Series Editor

Janusz Kacprzyk, Polish Academy of Sciences, Warsaw, Poland

The series "Studies in Big Data" (SBD) publishes new developments and advances in the various areas of Big Data- quickly and with a high quality. The intent is to cover the theory, research, development, and applications of Big Data, as embedded in the fields of engineering, computer science, physics, economics and life sciences. The books of the series refer to the analysis and understanding of large, complex, and/or distributed data sets generated from recent digital sources coming from sensors or other physical instruments as well as simulations, crowd sourcing, social networks or other internet transactions, such as emails or video click streams and other. The series contains monographs, lecture notes and edited volumes in Big Data spanning the areas of computational intelligence including neural networks, evolutionary computation, soft computing, fuzzy systems, as well as artificial intelligence, data mining, modern statistics and Operations research, as well as self-organizing systems. Of particular value to both the contributors and the readership are the short publication timeframe and the world-wide distribution, which enable both wide and rapid dissemination of research output.

The books of this series are reviewed in a single blind peer review process.

Indexed by SCOPUS, EI Compendex, SCIMAGO and zbMATH.

All books published in the series are submitted for consideration in Web of Science.

More information about this series at https://link.springer.com/bookseries/11970

Herwig Unger · Mario Kubek

The Autonomous Web

Herwig Unger
Lehrstuhl für Kommunikationsnetze
FernUniversität in Hagen
Hagen, Nordrhein-Westfalen, Germany

Mario Kubek
FernUniversität in Hagen
Universitätsstraße
Hagen, Germany

ISSN 2197-6503 ISSN 2197-6511 (electronic)
Studies in Big Data
ISBN 978-3-030-90938-3 ISBN 978-3-030-90936-9 (eBook)
https://doi.org/10.1007/978-3-030-90936-9

This Springer imprint is published by the registered company Springer Nature Switzerland AG
The registered company address is: Gewerbestrasse 11, 6330 Cham, Switzerland

Preface

The World Wide Web became the major information source for many people within the last years; some even call it the brain or lexicon of mankind. For any arising questions, any facts needed or any multimedia content wanted, a web page providing the respective information seems to exist. Likewise, it seems that sometimes there is nothing what has not been thought, written, painted or expressed in any other form before. In particular, for researchers and scientists, it is a fast deal now to retrieve information about the works of their colleagues and check whether an idea is new or already applied anywhere in this world. Geographic distances become more and more meaningless because of this rapid exchange of information, and the world appears somehow like a village.

Despite this still permanently growing amount of available information, its management still did not change much over the past decades. Big search engines, and mostly Google as a monopolist with over 90% of usage, are needed to map an information request to the right location (Uniform Resource Locator, URL), where the respective information is found. Although hardware and search engine technology has rapidly improved, the major principles are still the same. Web pages are downloaded (copied), indexed and their main keywords or contents are stored in huge (centralised and not always continuously updated) databases. This big amount of data can be accessed by formulating a request with a few, well-selected keywords, which will result in an answer in form of a plain ASCII list with links to matching contents and sometimes containing a set of pictures and videos with fitting contents. Besides that inopportune user interaction, the current approaches of web search lead to a high concentration of too many sensitive information at a few, huge search engine companies and transform the users into transparent ones, who are predictable in most of their activities and aims.

Consequently, the need for any new technologies for the autonomous self-management, more timely information handling and the user's interaction with it in the Web arises. From the authors' point of view, this problem is mostly a navigational one for which a lot of analogues are known. Amazingly, the most common one is the management and navigation in the street system of a country. Similar to the Web, nobody is able to oversee, know or control the status of all highways, streets and places in a country, but a person may know only her/his local neighbourhoods quite well and is thus able to navigate perfectly in it. For any far-distance navigation to places managed by other people (regions), only a rough orientation might be possible. A global signpost system and a few global maps can help to find the right directions there, while local people (knowledge) may help to orientate within the conditions of another city. Considering the main principles of the management and functionality of a traffic system, namely the independent, local management as well as the use of local knowledge in the respective situation, those aspects are the basis for a perfectly working system. The application of these considerations in the World Wide Web means to also remember and favour decentralised systems' working principles: in fact, if the original approach of the Web is considered, web servers with their linked and locally managed documents build exactly such a system but lack any real and automated navigation possibilities. Furthermore, there are no efficient mechanisms to group similar or related information in a (maybe hierarchic) manner and make it accessible to the human's top-down or bottom-up thinking.

At this point, the interested reader may ask her- or himself, which organisational and functional principles derived from nature and especially the human brain may be adapted to realise the desired, new working principles of the Web. Indeed, the authors of the contributions in this volume were greatly inspired by Einstein's saying 'Look deep into the nature, then you will understand everything better'. Therefore, many of the presented methods are inspired by nature as well as physical or biological counterparts and are realised within a computer systems' environment; the major idea, the text-representing centroid terms used for categorisation purposes were derived as a special form of generalised medians from the center of mass in physics. Most other approaches follow the strict natural design principle of locality, i.e. work without overseeing the whole set of data and structures and work with locally available information, only.

To explain this new point of view on the Web, the book is structured as follows:

- In Chap. 1, we briefly review the structure and development of the World Wide Web, while special attention is given to the search and information management in it.
- Chapter 2 discusses an approach for a new kind of locally working (medical) recommender system, which in particular aims to summarise and repeat the major ideas of the co-occurrence-based text mining and of course of the mentioned text-representing centroids.
- Based on this knowledge, Chaps. 3–5 present three versions of a new, fully decentralised and integrated web search engine, which realise the imaginations of the authors and their co-workers to create a new kind of an underlying information management for the Web. Another link-optimising information system is presented in Chap. 6 as an alternative for medium-sized, web-based information systems.
- Finally, Chaps. 7–11 explain the major but usually application-specific and not easily accessible methods of our new information management approach. Due to its general importance with respect to grouping and ordering information, two

chapters are addressing problems of clustering: after a survey in Chap. 7, a locally working clustering algorithm is introduced in Chap. 8. Chapter 9 deals with a special aspect of decentralised routing with incomplete information. Last but not the least, Chaps. 10 and 11 will present important aspects of modern knowledge management: finding frequently used sequences and applying a time-depending information management (e.g. oblivion).

At this point, the editors want to thank all researchers of the 'Team of Communication Networks' for their effort and for following and developing our maybe strangely sounding, first ideas over many years. As a result, many working applications were implemented as well as publications and finally this book was written. Therefore, in each of the chapters, the authors present their recent work results in a short, condensed manner.

All of us wish our readers inspiring hours with this book, and we will be grateful for any responses and opportunities to present our works to the community of our much-valued colleagues all around the world.

Hagen, Germany
August 2021

Herwig Unger
Mario Kubek

Acknowledgements

Both researchers are lucky that after the first meeting 15 years ago, they were able to identify common research interests and, regardless of initial difficulties, found a common language, common research goals as well as a common vision for their work. In 2010, this resulted in the opportunity to research and teach together in a team, which is working together for over 11 years now. Supported by many colleagues and students at all levels, some of them from our permanent research partner King Mongkut's University of Technology North Bangkok (Thailand), those dreams could be realised step by step. Many inspiring moments, weeks of hard work followed by happiness or deep depressions as well as difficulties of various kinds to solve were on our way. This book contains the summary and the reminiscences of what two researchers and their team may achieve, if they feel the passion for research and follow their vision. On the way, we enjoyed every single minute, were thankful for every, often unexpected, help obtained and feel deeply sorry that this close, joint work might have to come to an end very soon.

Contents

About the Authors

Prof. Dr.-Ing. habil. Dr. h.c. Herwig Unger *1966 (right in photo) received his Ph.D. with a work on Petri Net transformation in 1994 from the Ilmenau University of Technology and his doctorate (habilitation) with a work on a fully decentralised web operating systems from the University of Rostock in 2000. Since 2006, he is a Full Professor at the FernUniversität in Hagen and the head of the Department of Communication Networks. In 2019, he obtained a honorary Ph.D. in Information Technology from the King Mongkut's University of Technology in North Bangkok (Thailand). His research interests are in decentralised systems and self-organization, natural language processing, Big Data as well as large-scale simulations.

He has authored more than 150 publications in refereed journals and conferences, published or edited more than 30 books and gave over 35 invited talks and lectures in 12 countries. Besides various industrial cooperations, e.g. with Airbus Industries, he has been a guest researcher/professor at the ICSI Berkeley (USA), the University of Leipzig and other universities in Canada, Mexico and Thailand.

PD Dr.-Ing. habil Mario Kubek *1979 (left in photo) is a Senior Researcher at the FernUniversität in Hagen, from which he received his Ph.D. with a thesis on locally working agents to improve the search for web documents in 2012 and his habilitation with a work on how to create a fully integrated, decentralised and librarian-inspired web search engine in 2018. For this work, he was awarded the Annual Prize of the Faculty of Mathematics and Computer Science at the FernUniversität in Hagen for the best scientific work in 2018.
His research focus is on natural language processing, text mining and semantic information retrieval in large distributed systems. He is also interested in mobile computing environments, contextual information processing and secure software engineering.

Chapter 1
State-of-the-Art Survey on Web Search

Georg P. Roßrucker

1.1 Introduction

With respect to the concept of a decentralized *WebEngine 3.0*, also referred to as *WebMap*, laid out in [1], this paper intends to present an in-depth and state-of-the-art study of literature and related research. From this, a classification of the web is derived, that on the one hand highlights the structure of the web and content distribution, and on the other hand, depicts relevant aspects of web search. This will then serve as a foundation for the later system design, and development of algorithms for the proposed *WebMap*.

Section 1.2 puts the focus on research regarding the structure of the web. All searchable content is embedded into this structure, which determines the scope of web search. Section 1.3 presents a taxonomy of web search, deriving aspects by which *web search* can be differentiated: User intention, search subject, and use-cases. Section 1.4 describes the supporting pillars of web search and associated concepts and technologies, concerning content discovery and natural language text processing. Section 1.5 presents challenges of web search, such as deriving suitable results and determining relevance and ranking measures. In addition, approaches of decentralized search are discussed. Lastly, Sect. 1.6 concludes with a classification of the web and recommendations regarding further research.

1.2 The Web

1.2.1 Structure of the Web

Size of the web

The world wide web was invented by Tim Berners-Lee, who published the first website[1] in 1989. In [2] he documented the features, that *the Web* should provide,

[1] The birth of the Web: https://home.cern/science/computing/birth-web.

H. Unger and M. Kubek, *The Autonomous Web*, Studies in Big Data 101,
https://doi.org/10.1007/978-3-030-90936-9_1

and by which means this shall be achieved: Naming schemes, network protocols, and data formats, such as hypertext and hyperlinks, allowing to reference information by mutual, or one-way linking. Since Berners-Lee founded the web, it has expanded at high speed: Available data show, that the number of websites has grown exponentially until it reached approximately 650.000 in 1997 [3]. Today, the number of websites approaches two billion [4]. Both numbers refer to the count of unique host names, from which it can be concluded, that *websites* consist of multiple *web pages*, i.e., documents, sharing the same host name. An estimation based on indexed web pages [5] suggests, that there exist at least 5.36 billion web pages as of today.[2]

Graph structure of the web

The referencing scheme, proposed by Berners-Lee allows connecting related content on the web. It enables users and search engines to navigate from one page to another. These *Hyperlink References*, consist of a descriptive text and a URL, which addresses the particular resource on the WWW. Besides others, a URL contains a host identifier and specifies the port, protocol, and location of the resource.[3] The main technology behind the resolution of a host identifier, usually a domain name, to the host's IP address is DNS.[4]

Given the hyperlink references, which connect resources across different hosts, the web forms a decentralized graph structure, in which the vertices are web pages, and hyperlinks represent the edges between them. A study presented in [6] found, that this graph is divided into four major sections:

1. The *strongly connected core (SCC)*, in which websites are strongly and mutually connected. The mutual links build a strong network structure.
2. The *OUT section*, consisting of nodes referenced by the SCC but not referencing back. The unidirectional links form a tree structure.
3. The *IN section*, consisting of nodes referencing the SCC but not being referenced back. The unidirectional links form an (inverse) tree structure.
4. The *disconnected section(s)*, consisting of nodes, i.e., websites that are not part of the graph and cannot be reached.

Figure 1.1 illustrates this structure, which is also referred to as the *bow-tie structure* of the web.

Repeated studies [7, 8] confirmed the existence of a strongly connected core section. Yet, the newer findings suggest, that the proportions of the four sections should be revalidated, since they differ strongly, depending on the crawling process, and seem not to be structural properties of the web graph (anymore) as suggested in [6]. Meusel et al. [7] used a dataset of 2.5 billion nodes, compared to the 200 million nodes of the former study, [6].

---

[2] Number of indexed pages as of February 5th, 2021: https://www.worldwidewebsize.com/.

[3] URLs, a subtype of URIs, are defined as per RFC2986:
https://datatracker.ietf.org/doc/html/rfc3986.

[4] DNS is defined as per RFC1035: https://datatracker.ietf.org/doc/html/rfc1035.

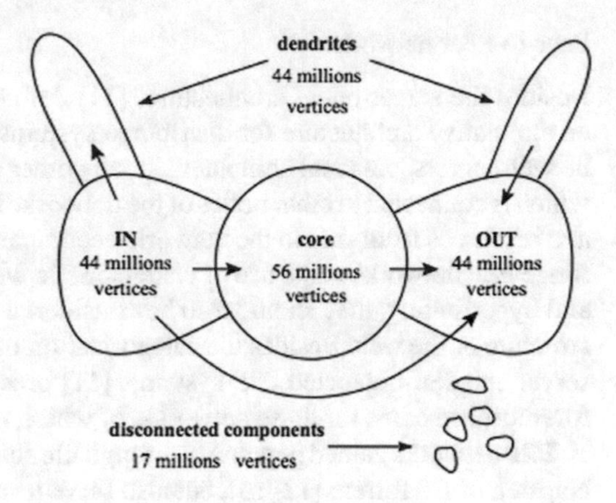

Fig. 1.1 Bow-tie structure of the web. *Source* Wikipedia, based on [6]

Hidden, shallow, deep, and semantic web

The *hidden web* refers to the disconnected sections of the web, which cannot be crawled and discovered by web search engines, that simply follow the existing link structures.

For the connected sections, a distinction between the *shallow web* and the *deep web* can be made. The shallow web refers to static web pages, that do not change their appearance over time. The deep web in contrast refers to dynamic web pages, that render different outputs for a given URL for repeated requests, depending on individual users, or input data, e.g., pages hidden behind web forms. In [9] was shown that in the year 2000 the available information on the deep web exceeded those on the shallow web by at least 400 times.

Lastly, the *semantic web* and the Resource Description Framework (RDF) should be mentioned. RDF serves as an extension to the web, introduced in [10]. It allows the addition of contextual information to web content and to build an expressive knowledge graph, consisting of vertices, i.e., entity-representing URIs, and edges describing the relationship between them. This technology extends web pages, i.e., hypertext elements, by machine-readable descriptive and relational information in addition to regular hyperlinks between resources.

1.2.2 Early Approaches to the Use of the Web

The graph structure of the web reflects the server-side of its distributed server-client architecture, in which web servers provide websites, or other web services, to clients, who navigate through the hyperlinked environment and consume these services and contents with client tools, such as web browsers.

Peer-to-Peer networks

Besides the server-client architecture, [11] defines *peer-to-peer (P2P) networks* as an alternative architecture for distributed systems, in which each participant, may it be web servers, personal computers, or any other device act as an autonomous node, which is connected to other nodes of the network. They provide and consume services and resources from, and to the network, accompanied by a set of protocols and rules. Since P2P networks make use of characteristic web technologies, such as hypertext, and hyperlinking, they should also be considered in an comprehensive survey of the structure of the web. Besides the categorization of decentralized systems into client-server and self-organized P2P systems, [11] presented the main challenges and the foremost use cases for these networks, of which various applications have emerged.

P2P networks gained popularity through file sharing applications, such as Gnutella, Napster, or BitTorrent [12–14], but also serve a variety of other use cases. News and collaboration networks, such as Usenet, or Freenet [15, 16], allow decentralized communication, and distribution of information, e.g., to prevent censorship, or other external actions of influence, or suppression. The web of trust [17] is a P2P network intending to build a network of transitive trust, in which the trustworthiness of entities can be derived from, and verified through mutual trust relations, safeguarded by cryptography. More modern use cases are cryptocurrencies, such as Bitcoin [18], which rely on P2P networks for the management and distribution of transactions and the underlying blockchain, protecting its integrity from unauthorized modification.

Web Operating Systems

The concept of *Web Operating Systems (WebOS, WOS)* goes beyond the usually unique purposes of P2P applications, presented above. [19, 20] present general frameworks for P2P-based infrastructures, that support a variety of applications, regardless of their use case.

The *WebOS framework* [19], proposes an abstraction layer to distribute functionalities, usually provided by a local computer's operating system. Distributed data storage, remote execution, access and security mechanisms, resource management, and a global naming scheme are presented, as core components of the framework. This allows application developers to focus on functionality, rather than the underlying infrastructure and implications concerning scalability, availability, computing power, content distribution, or load balancing when building distributed applications.

Since, among others, it reduces the complexity of developing distributed applications, this model of abstraction has reached widespread acceptance across web and application development today. Popular tools, such as Kubernetes,[5] provide functionality allowing to orchestrate, scale and distribute applications to on-demand needs. However, providers of such services, for example, Amazon Web Services, or Microsoft Azure, are not operating on public P2P networks but set up their own distributed infrastructures for abstraction and virtualization.

[5] Documentation and overview of Kubernetes: https://kubernetes.io/docs/concepts/overview/what-is-kubernetes/.

The *WOS framework* presented in [20, 21], aims at similar use cases, i.e., sharing resources and capacities across the nodes of an underlying P2P network, to conduct tasks that a host, limited to its resources cannot perform. The proposal is based on an incremental, version-based approach, in which nodes provide OS features, such as processing power, disk storage, memory, and software execution as a service. Therefore, each participating peer provides a catalog of services and supported versions it is capable to provide. A common communication protocol, also versioned, allows peers to pass on requests and responses among each other. Since there is no central management or central catalog of resources, the network is highly adaptable and each peer is eligible to participate in already existing services or provide modified, or individual services to the network. Peers providing defined versions of services can form communities, so tasks and load can be distributed, and balanced efficiently across them. Finally, the concept explicitly proposes the construction of a distributed web search engine as exemplary service.

The distributed OS approaches have in common, that they rely on remote execution of code and software. Specifically for dynamic web pages, the *Common Gateway Interface* (CGI) allows web servers to execute downstream applications on the host.[6] For example, PHP, Perl, or any other server-side software may be invoked to render an HTML page on-demand and with respect to user input, which is usually triggered by users retrieving the web page through their web browser. While CGI serves as a web server-specific code execution gateway, *Remote Procedure Calls* (RPC), and *Distributed Object Communication* (DOC) extend the scope of remote execution to any other software [22, 23]. RPC and DOC allow the distribution of functions and objects to other computers where they can be invoked by a connected peer. After execution, function result, or object reference are returned to the client. The distributivity of functions and objects needs to be considered when designing a distributed application. There also need to be protocols and frameworks in place, that govern distribution, execution, and communication, like Java RMI, or specific P2P protocols in the case of decentralized applications.

1.3 Taxonomy of Web Search

Web search can be adapted differently depending on *user intention*, *environment*, and *use case*. A taxonomy of web search is elaborated in this chapter.

1.3.1 Intention

Given the graph structure of the web and the continuing increase of available content, questions regarding efficient web navigation arise. According to [24], the fundamen-

[6] CGI is defined as per RFC3875: https://datatracker.ietf.org/doc/html/rfc3875.

tal questions of web users regarding navigation are: "*Where am I?*", "*Where can I go?*", and most importantly, "*How do I get where I want to go?*". The last question can be separated into a technological part "*How do I get ...*" and an intentional part "*... where I want to go*". The technological part of this question can be answered with the fundamental web tools, namely *web browsers*, *web page design*, *specialized client and server-side tools*, and *web search engines*. The intentional part has been the subject of user behavior studies looking into the actual needs of users.

In [25], the intentions of users, derived from search queries, were categorized as follows:

1. *Navigational*: Users know where they want to go but do not know the specific address/location;
2. *Informational*: Users want to find out more information about a specific topic;
3. *Transactional*: Users look for specific web services, such as online shopping, downloads, or similar. It was also shown that almost 50% of search queries were in the informational domain.

Jansen et al. [26] confirmed this categorization and added, that web search engines are primarily used to satisfy informational needs. 80% of user queries could be assigned to the informational domain, and only 10% each, to the navigational and transactional domain. Figure 1.4 illustrates the intention of web search and the underlying objectives presented in [27].

In [28], another framework for automated classification of web search queries was introduced. It distinguishes between informational and not informational requests, where not informational comprises of, e.g., transactional, and ambiguous queries. It was shown, that, despite the predominant informational use of search, the classification of queries to either of the two domains differs strongly with regard to the contextual categories, that they were assigned to. While queries in the contextual categories of *news* and *science* were strongly associated with the informational domain, queries in the categories of *games* and *adult content* were clearly with the not informational, i.e., transactional domain.

Finally, from [24] it can also be concluded, that different technologies, have their strengths in satisfying different user intentions. As the following list depicts, web search engines are the main tool to address the informational needs:

- **Web browsers → Navigational needs**
  Site visitation and re-visitation, navigating backward, forward, entering URLs, bookmarking, etc.
- **Web page design → Navigational and transactional needs**
  The web page design primarily aims to serve user-friendly navigation and if applicable the completion of transactions on a specific website.
- **Web search engines → Informational needs**
  Providing suitable information resources derived from a search engine's index for a given search query.
- **Specialized client and server-side tools → Multi-purpose**
  Browser plugins, server extensions, and website features, that may serve any of a user's needs.

1.3.2 Subject: Open Versus Closed Domain Search

The domain classification helps to differentiate web search with respect to the search space. The distinction is important due to different network structures, in terms of size, link structures, users, and content.

Open domain web search

Refers to information retrieval on the broader Internet. As elaborated in Sect. 1.2, information distribution on the Internet occurs uniformly in terms of format (hypertext) and linking (hyperlinks). Yet, information is provided unstructured, maintained decentralized, and linking has grown organically and unsupervised. Therefore, classic information retrieval strategies in the open domain, make use of the graph structure and natural language processing such as presented in [29–31]. The main challenges of search in the open domain come with the amount of available data, i.e., scalability, appropriate filtering of relevant information, and covering all available content from the shallow to the deep web.

Closed domain web search

Refers to search in segregated network structures, such as enterprise Intranets, social media platforms, or single websites. Major challenges of information retrieval specific to enterprise Intranets were discussed in [32]. Compared to the open and mainly hypertext- and hyperlink-based Internet, closed networks often provide unlinked information of various formats in structured and unstructured repositories. They range, among others, from hypertext sources like external and internal websites to documents on file shares, specialized applications and databases, email, and people information. To ensure a need-to-know regime imposed by data classification and legal requirements, information is often protected with user-level access control mechanisms. Therefore, organization- and job-specific search problems along with the complexity of the search space require individual search strategies for the specific domain.

Further research supports the assumption, that efficient search in closed networks is a highly individual matter. Besides various approaches proposed to improve closed domain search [27, 33–37], it can be summarized, that the following aspects should be considered, when implementing a search engine in a closed domain environment:

- Continuously assessing search scenarios, and user behavior.
- Continuously identifying and integrating (heterogeneous) data sources, i.e. repositories and adapting to changing environments.
- Data pre-processing and information extraction, tailored to organization-specific search scenarios, e.g., mapping of data to intentions, and other means of building a suitable search index.
- Only providing information complying with a set of rules, respecting classification, legal requirements, and other restrictions.

Besides multi-repository Intranets, strategies known from search on the broader Internet could be applied to retrieve information from closed domain hypertext environ-

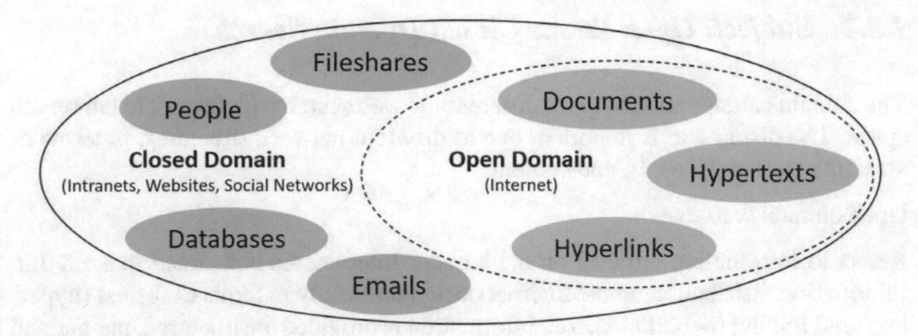

Fig. 1.2 Open versus closed domain web search. *Source* Own illustration

ments, such as personal or organization websites, hypertext-based Intranets, and similar. However, [38] states that in these closed environments the search performance of link-based search techniques is relatively poor compared to their application on the Internet, due to the smaller size, structure, and contextual focus of the collections. They propose generating additional links based on observed user behavior to improve search performance.

It can be concluded, that in general, search in the open domain, i.e., the Internet, is not applicable to closed domains, or vice versa. Parallels regarding hypertext search exist but search engines need to be tailored to the specific use cases and search scenarios in the respective environments. Figure 1.2 depicts potential search spaces of the open and closed domains. Considering that there exist uncountable, individual closed domain environments, there cannot be a one-fits-them-all search solution.

1.3.3 Use Case: Query Versus Content-Based Search

This section introduces another classification of web search based on the origin of the search requests. The results of a web search can in general be interpreted as link recommendations for a given request, which may have different origins. They can be user-formulated search queries, or data and content derived from documents like websites.

Query-based web search

Users act actively, i.e., asking a specific question by consciously entering a search query and expecting matching results. Analyzing these human-formulated search queries and deriving the user's intent, as discussed in Sect. 1.3.1, is, therefore, an important task to improve search quality. In [39], several such approaches were evaluated, like the specificity of a query, which serves a measure that can be matched with the search results' level of specificity to pick the right results depending on the request. Propositions regarding the specificity were made in [40]. In [41] the evolution of search queries in interaction with continuously adapting search engines was analyzed from a linguistic perspective, stating that from several aspects (functionality, structure, and evolution), search queries share properties of a protolanguage. This observation may be relevant in the further development and improvement of search engines' capabilities to process and analyze queries.

Content-based web search

Users act passively, consuming links presented to them by recommender systems, usually depending on their present location and context, e.g., a product in an online shop, a video clip, or any arbitrary web page, or document. In general recommender systems aim to anticipate a user's intention and provide guidance towards the ideal path of navigation. More sophisticated systems take additional data such as user profiles into account, or utilize feedback mechanisms to refine future recommendations [42]. An approach of link recommendation based on user similarity and previously recorded web traverses was presented in [43]. Recommender systems gained popularity in the e-commerce and online entertainment sector, where they are used to promote and reference products, or online media [44, 45]. Often recommendation systems apply collaborative filtering to infer the interests of single users based on the common interest of an associated reference group [46, 47].

1.4 Supporting Pillars of Web Search

In order to build a search engine, it is necessary to create an index of all the available content, process it, and annotate it, so that the search algorithm is capable of delivering the best-fitted results for incoming search requests. In retrospect of the previously introduced taxonomy of search, the focus of this chapter is on open domain search only. The following aspects of web search are covered:

- Content discovery: Crawling the web for available resources
- Content processing: Text pre-processing, keyword extraction
- Text mining: Deriving information from documents and the corpus

1.4.1 Content Discovery

The component discovering new resources on the web, i.e., collecting web pages' URLs, and downloading their content, is known by various names, such as *crawler*, *spider*, *indexer* or *search robot*. As web content is continuously growing and changing, the efficiency of a crawler in capturing an up-to-date snapshot of the web is crucial for the competitiveness of a search engine.

Some of the main challenges of effective crawling were presented in [48]. The *continuous growth of the web* needs to be addressed, e.g., by prioritized crawling, to identify important pages fast. Adjoining, techniques to *minimize network load and communication overhead* should be considered in the crawler setup. In order to maintain a *fresh and competitive index*, establishing a revisiting/refresh policy is essential, to *observe changes* over time.

Depending on the objective of a search engine, filtering of content should be reflected by the crawling strategy. According to [49, 50], the main strategies are *focused*, also known as *preferential crawling*, and *blind/general purpose crawling*. While the first focuses on pages of a given context only, i.e., discards all pages not matching a predefined pattern, the latter crawls all sites reachable from a given starting point regardless of any side conditions. In addition, *ad-hoc crawling* refers to a user-centric crawling strategy that only considers links, relevant for a given user in an on-demand process.

As discussed in Sect. 1.2.1, [9] and others [50, 51] claim that a large share of useful information on the web is hidden behind dynamic websites, e.g., only accessible through web forms and dynamic websites. Hernández et al. [50] surveyed 21 crawling approaches, proposed between 1998 and 2015, that go beyond the trivial task of following links, aiming at information that is hidden in the deep web. First of all, they found out that there exist no benchmarks, or standards to compare or measure the performance of crawlers. Therefore, they proposed a comparison framework comprising of aspects such as automatic form detection and form-filling capabilities. Yet, their results revealed, that crawlers perform differently well, with respect to the context and underlying crawling strategy. Also, most approaches are not fit for future innovations, suggesting that for example, modern crawlers need to consider web technologies, such as javascript-based websites, dynamically being rendered by the client software, altering the statically hyperlinked appearance of the web. This leaves some open questions for future research and development.

A website operator can utilize several technologies to support efficient web crawling and perform search engine optimization (SEO). Concerning web crawling, a robots.txt as defined in the *Robots Exclusion Standard*[7] can be placed in the root folder of a domain, giving guidance on which parts of the website should, or should not be crawled, or archived. Additionally, the location of a sitemap can be specified. The *sitemap*[8] reveals the internal structure of a website, allowing for a structured crawling approach and supporting link detection.

The content discovered by crawlers needs to be processed for further use in search engines. This includes pre-processing, and alignment of natural language texts, keyword extraction, and extraction of information for classification. This is referred to as

---

[7] Robots.txt Specifications https://developers.google.com/search/docs/advanced/robots/robots_txt.

[8] The Sitemap Protocol https://www.sitemaps.org/protocol.html.

natural language processing (NLP). Some techniques need already be applied during the crawling process, e.g., to identify the next crawling target during a focused crawl.

1.4.2 Text Processing and Keyword Extraction

After discovering and identifying documents, e.g., web pages, and before applying *text mining* techniques such as categorization, sentiment, or co-occurrence analysis an adequate pre-processing of the document corpus is required. Various techniques exist and were presented in [52, 53]. Figure 1.3 outlines the general pre-processing workflow.

Extraction

The first step of the pre-processing workflow is to extract text from a given document [53]. In the context of the web this is the process of extracting plain text from the HTML structure of web pages, but can also be applied to any other data format. The extracted text is then tokenized, i.e., single words are derived and provided as a list for subsequent steps.

Due to its complexity, further steps are required to simplify and align natural language texts, so that they get processable and comparable by machines regardless of, e.g., grammar, or writing style. *Stopword removal* and *Stemming* are two examples. However, it must be kept in mind, that simplification comes with a loss of information and should, therefore, always be aligned with the capabilities of later text mining and classification methods.

Stopword removal (SWR)

Stopwords are language-specific features, of decorative or grammatical nature, and are shared across all texts in a given language. SWR intends to remove any of those words, which are irrelevant in terms of content and expressiveness of a given text. Ladani and Desai [54] presents a survey on stopword identification techniques, stating that SWR can reduce the amount of data to be processed by 35 up to 45% and is, therefore, an important method to improve the performance of subsequent tasks.

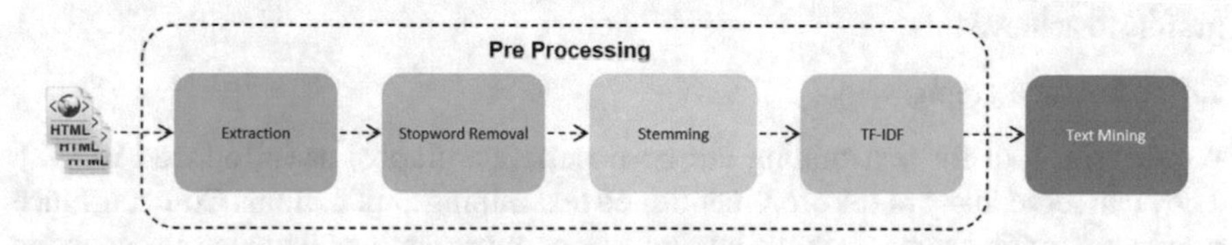

Fig. 1.3 Pre-processing workflow for a given document corpus. *Source* Own illustration

Stemming

Grammar in various languages makes words subject to syntactical changes. Examples are the declension of words by number (plural, singular), case, or gender, and conjugation of verbs, e.g., by person, or tense. In addition, there exist different words, e.g., nouns, verbs or adjectives sharing a common base form: Clusters of words with the same *stem*. To create a comparable set of terms across multiple documents, the process of stemming can be applied. Stemming reduces the words of a given text to their base form, and by doing so, eliminates grammatical differences, and normalizes terms across documents. A survey on several stemming techniques was presented in [55].

TF.IDF

The *term frequency-inverse document frequency* (TF.IDF) is a statistical measure indicating important and characterizing terms of a document in a text corpus. It puts the frequency of a word in a given document in relation to the inverse ratio of the word's appearance in all documents of the corpus. The more occurrences of a term in a specific document and the fewer occurrences in other documents, the higher the characteristic property of this term with respect to the document [56]. By this method, characterizing keywords can be derived, and non-characteristic words (like stopwords) appearing frequently across all documents can be identified and eliminated. Following [52], the TF.IDF measure can be used as a weighting property for terms in subsequent text mining tasks. They also propose document frequency thresholding to remove infrequent words from a text. The assumption is that terms, which do not occur in a minimal proportion of the corpus are unimportant and not suitable for further classification.

1.4.3 Text Mining

Given a pre-processed text corpus, the next step in constructing a web search engine is to extract information, that supports the later search algorithms. The objective of *text mining* is to further analyze documents with respect to keyword extraction, contextual classification and clustering, specificity, similarity, and sentiment. This section will give an overview of methods that are mainly based on co-occurrence graphs to achieve this.

Co-occurrence graphs

A powerful tool for text mining are co-occurrence graphs, as introduced in [57]. They can serve many of the aforementioned text mining applications. Co-occurrence graphs are graph databases that contain vertices, representing all the terms occurring in a given text corpus. They are created by successively processing all documents of a text corpus and adding new words to the graph. If two words co-occur, i.e., were observed in the same frame, e.g., a sentence, or a paragraph, an edge is created to link the corresponding vertices. The *weight* of an edge can be determined by counting the frequency at which the two words occurred together in the text corpus. Its inverse can be interpreted as a measure of the *distance* of the terms.

Keyword extraction

In analogy to the geometric center of mass, [58] proposed an algorithm that computes a single representing term for a given document, the *text representing centroid* (TRC). It is derived from a co-occurrence graph by identifying the vertex, i.e., term, with minimum average distance to all words of the given document. This characteristic term can be used for *contextual classification* or as a *keyword* for the document. More keywords can also be derived by extracting alternative TRCs from the neighborhood of the TRC.

Mihalcea and Textrank [59] proposed *TextRank* as a method to extract important words and sentences from a document. The method adapts the concept of the *PageRank* [30] and is based on a document-level co-occurrence graph. It either reflects word relations, based on their co-occurrence, or sentence relations, based on common features, like bag-of-words or other measures. All words or sentences receive an initial score and an iterative voting mechanism between the connected vertices is applied until convergence is reached. Finally, the vertices can be sorted by their score and the highest-ranking can be derived as document representatives.

More methods of term and keyword extraction, aiming to project a given document to a single, or few keywords, like the aforementioned TD.IDF statistics were presented in a comprehensive survey in [60].

Contextual classification

Assuming that similar documents have TRCs in relatively close distance, clustering terms on the co-occurrence graph may help to classify documents by a joint context. Therefore, methods to determine clusters on the co-occurrence graph are needed to identify groups of words representing a common context. Simcharoen et al. [61] presented an approach for clustering, also utilizing the centroid approach. Here, each cluster is identified by a cluster-centroid term, surrounded by all terms of a given maximum distance. New terms are assigned to clusters, while clusters and their identifying centroid terms adjust over this iterative process until an equilibrium state is achieved. A broader and comprehensive survey on graph clustering methods was presented in [62].

Similarity

Besides contextual classification, co-occurrence graphs and TRCs can be utilized to measure the similarity of documents. Koutra et al. [63] surveys graph-based approaches, such as *subgraph matching*. In this case, the similarity of two documents can be derived from mutual features of their subgraphs derived from a common co-occurrence graph. Kubek and Unger [58] suggests deriving the similarity of two documents by the distance of their TRCs on the co-occurrence graph, the closer they are, the more similar the documents are. Their results show that this method allows the determination of similarity for documents, even if a different lingo is used. This is possible because a mature graph contains the words of all inserted documents, and words of the same context are assumed to be closely connected, even if they did not occur in all similar documents. This gives an advantage over other methods, based on bag-of-words, such as cosine similarity.

An alternative to the co-occurrence graph-based similarity measures is *Doc2Vec*, which was introduced in [64]. It derives a vector representation for combinations of words in sentences, paragraphs, or documents. It is based on the concept of *Word2Vec* [65], which introduced vector representations for single words. Comparing these vectors allows determining a semantic similarity score for single terms, sentences, or whole documents, regardless of the bag-of-words.

Specificity

Contextual classification by TRC, cluster, or similarity has no expressiveness about the *specificity* of a document, as it does not account for the spread and diversity of words in a given document. Kubek et al. [66] defined the maximum of the word-TRC-distances as a measure of *diversity*, indicating the specificity of a document or query relative to others. The lower the distance, the more specific a document (and the more general, otherwise). An alternative approach could be, to measure the variance of the distribution of word-TRC-distances of a document. In both cases, the interpretation follows the logic, that if the words of a document A are more narrowly (more broadly) distributed around its TRC, relative to another document B, A is more specific (more general) than B.

Sentiment analysis

Another field of text mining is *sentiment analysis*, intending to capture aspects such as tenor, mood, favorability, or contradiction with a specific topic. The main challenges are to distinguish *sentiment expressions* and their *subjects* in a given text. Additionally, suitable *sentiment classifications* are required to derive a quantitative measure. Given this, sentiments for more complex expressions, such as headlines, sentences, sections, or whole documents can be determined. Tang et al. [67] presents a comprehensive survey on subjective classifiers, sentiment detection, and classification. Pang et al. [68] proposes a classification approach by statistical means based on a manually recorded training set and machine learning techniques. Nasukawa and Yi [69] applied a syntactic parser to identify the semantic relationships, and a sentiment lexicon to determine the sentiment towards a specific subject. They found that most of the failures of their application were caused by the complex structures of natural language texts. Kennedy and Inkpen [70] introduced a rule-based approach. Besides sentiment expressions also valence shifters are identified that affect the polarity of an expression, i.e., negate, intensify, or diminish the polarity of the respective sentiment. Approaches based on co-occurrence graphs were not found at the time of this writing. A first step could be to apply sentiment lexicons to the graph in combination with suitable algorithms, to evaluate the sentiment of a given text.

1.5 Search Engines

1.5.1 Challenges of Web Search

A search engine or recommender system needs to provide suitable results, for an input query (or document). The main challenges are to identify matching resources from the indexed dataset on-demand and rank the results according to their *relevance*. This facilitates an optimization problem, i.e., deriving a set of relevant pages, or a single best result, with respect to the input data. A trivial approach would be to return all documents that contain the search request terms, and match refining query operators, or regular expressions, to a given degree, and use this as a score for ranking.

Fish Search, presented in [71], is a client-based search engine, that iteratively crawls the web starting from a given document, processes subsequent documents by the means of aforementioned query mapping, and derives a score, by which the pages are then sorted and returned to the user. An improvement, *Shark Search*, was presented in [72], which all in front, made use of *relevance propagation*, i.e., a document passing on its relevance score towards linked documents, whose relevance is damped or boosted, accordingly.

1.5.2 Relevance Propagation

When multiple documents match the search request to a high degree, e.g., if the search for a general topic is applied on a pre-processed web index, the problem of sorting the results becomes even more imperative. Examples of approaches to solve this are the HITS algorithm presented in [29], and the PageRank in [30], which both make use of the web's underlying hyperlink structure to conduct relevance propagation.

The *HITS algorithm* [29] allows on-demand evaluation of the *authority* of a web page with regard to a given term, or topic, derived from the search request. It uses a *base set* of web pages, generated by expanding a *root set*, which was derived from a preceding text-based search. From the base set, *authority scores* are derived by iteratively passing on the scores via the underlying link structure, until an equilibrium state is achieved. The base set can then be sorted by the scores and returned as the search result. The *PageRank* score for web pages is computed similarly, but in contrast to HITS, does so for the entire web, based on a web graph, regardless of topic, or a pre-requested list of documents [30, 73]. During calculation, the PageRank score is also being passed on via outbound hyperlinks to other websites, until an equilibrium is achieved. The disadvantages of both approaches are that they require high computing power, and the scores need to be calculated on-demand (HITS) or need to be updated regularly (PageRank) to reflect the continuously growing and altering structure of the web.

Qin et al. [74] conducted a comparative study of relevance propagation and presented a generic propagation framework derived from previous research, such as [75, 76]. They compared *hyperlink-based* to *sitemap-based* and *term-level* to *score-level* propagation methods, stating that sitemap-based propagation performs better in terms of effectiveness and efficiency, while term and score-level do not significantly differ. The advantage of term-level propagation is, that it can be implemented in an offline search engine, as already seen with the HITS algorithm, whereas score-level propagation requires significant computational power and storage to compute a steady score, as in the PageRank.

Later, other approaches suggested incorporating user feedback and behavior into the computation of relevance measures. *Collaborative ranking*, introduced in [77], uses search result clicks as an indicator for relevance, promoting the respective document in subsequent searches. In [78], it was argued against, stating that click rates are not a measure of the actual relevance of a document, but only reflect the confidence in its relevance, induced by a high ranking. They showed this self-amplifying nature of collaborative ranking by analyzing search engine clickthrough logs, inferring the *real* document's relevance from the recorded user behavior. Furthermore, they proposed to predict the relevance of websites by feeding their relevance scores into a machine learning algorithm and applying it to unvisited websites.

Vassilvitskii and Brill [79] presented another approach to propagate relevance between web pages. They validated their hypotheses of relative relevance, i.e., that *relevant pages point to other relevant pages* and *irrelevant pages are pointed to by other irrelevant pages*. From this, they derived, that user feedback for one page can be propagated to pages with low distance on the web graph, without the need for further feedback or text analysis. This also brings the advantage, that low relevance pages, intending to improve their ranking by mutual linking (link farming), can easily be removed from the search results.

1.5.3 Query Analysis

Query analysis is key to understand the user's intent, as discussed earlier, and to derive and present a suitable set of search results at search time. The user intention is expressed by the search query and results in a classification of the search itself into *descriptive* or *associative* search. Descriptive search refers mainly to the informational needs, where the query *describes*, i.e., *points towards* the context of desired results. Associative search in contrast refers to navigational needs, i.e., locating resources, that match a given input topic. The query serves as a common denominator or a *contextual projection* of the desired resources.

Concerning associative search, query mapping and other approaches can be applied to identify documents, that exactly match the input terms and query operators. However, more advanced search engines would make use of more than just trivial query mapping, e.g., *contextual* matching. *One key search*, for example, should compile search results by ideally considering synonyms, or other contextually related terms and topics. For more complex queries, like the search for similar documents, contextual representations of the query (input document) need to be derived and mapped with those of the indexed resources. Here, the challenge lies in the reduction of a query, to derive a contextual representation. Kubek et al. [80] addresses this by hierarchies of TRCs, i.e., for a document, different levels of TRCs can be derived, as from sentence to section to the document level. Likewise, [81] proposes, to derive sub-topics, from a document. Based on these representations, subqueries can be generated and resulting resources receive relevance scores for each sub-topic.

The implementation of such query mapping and contextual approaches becomes increasingly complex with the continuous growth of the web and the indexed data. MapReduce [82] is a parallelized programming model, that was introduced to address efficient processing of tasks, too complex for a single computer, like processing and calculations based on search indexes or other big datasets. It is based on the functional programming operations, *map* (filtering, sorting, creating maps of key-value pairs) and *reduce* (aggregating information), but separates and shifts them into parallelizable processes, that can be executed on distributed computers. Key-value maps are the output of the *map* processes and input to the *reduce* processes. Therefore, they are maintained in shared storages, available to any of the distributed processes. They can be implemented in any distributed storage, like distributed hash tables, e.g., Chord [83].

The following two examples illustrate the MapReduce pattern in the context of web search: First, by deriving search results based on contextual relations, and second, by a ranking of search results based on search engine clickthrough logs.

To *derive search results*, there are two distributed *map* processes. The first derives a key-value set, consisting of context representing terms derived from indexed documents (e.g., TRCs / topics) and the documents' URLs: $term \rightarrow URL$. The second *map* process derives a key-value set of terms and their contextual relatives, e.g., based on a co-occurrence graph: $term \rightarrow related\ term$. The *reduce* process, conducted at search time, would aggregate a list of related terms for a given input query's contextual representation (e.g., query TRC), based on the $term \rightarrow related\ term$ map. This list is then used in a second *reduce* process, where URLs are aggregated from the $term \rightarrow URL$ map. The list of URLs is returned as an (unsorted) search result.

In order to *rank a list of search results* by popularity, MapReduce can be applied based on recorded clickthrough logs of a search engine (queries and URL clicked). The *map* process derives a $URL \rightarrow query$ set from the logs. The *reduce* process derives the total count of clicks per URL regardless of the query from this key-value map. Ranking is applied by descendingly sorting the results by the count of clicks.

1.5.4 Distributed Search

In contrast to the aforementioned centralized search approaches, collaborative approaches intend to outsource the search engine's components, introduced in Sect. 1.4, into a P2P network, and provide decentralized search and ranking algorithms, independent of a single provider, and sharing the computational cost and network load across the participants.

One of such search engines is *The WebEngine*, introduced in [84]. It is based on a P2P network, co-occurrence graphs, and TRC applications. Each participating web server maintains a co-occurrence graph representing its local documents and provides a search form to the users. The algorithm first derives the TRC of an incoming search query and then retrieves local documents, represented by the same (or alternative) TRC. By utilizing the underlying server-based P2P network, queries can also be forwarded to other peers in the network, to extend the search range, to retrieve more documents related to the query's TRC.

With *TheBrain*, the initial WebEngine evolved further and a global co-occurrence graph was introduced, that comprises all information stored in the individual local graphs, and is itself distributed across all participating peers [85]. As a result, the computation of TRC during a search operation is not limited to the local graph anymore, but reflects all peers' documents, and produces unified results across all participants, regardless of the peer's local contextual focus. Since peers can attach URLs of their local documents to the corresponding vertices on the global graph, all documents related to the query TRC can easily be retrieved by extracting their URLs from the corresponding global graph's vertex.

A cooperative approach for link recommendation, also based on co-occurrence graphs and TRC was presented in [43]. In contrast to the server-based WebEngine and TheBrain, it facilitates a client-based link recommendation mechanism, in which clients are connected via a P2P network. Each client records its web traverses and maintains a local co-occurrence graph reflecting the individual context and interests of a user. Link recommendations for a given web page are derived from traverses of similar users, based on their co-occurrence graph similarities. Additionally, the graphs' vertices could be enriched with hyperlinks to related documents, to derive results for search queries, like in the WebEngine.

The approaches have in common, that they dynamically compute link recommendations based on co-occurrence graphs. However, this is computationally expensive and relies on the availability of peers in the underlying P2P network. Additionally, links are discarded immediately, and search results and recommendations have to be recomputed for every request. A concept for further development of the WebEngine aims to eliminate on-demand tasks reliance in the P2P network and expensive global graph computations by introducing static cluster and shadow files, referred to as the *WebMap* [1].

Other P2P search engines, such as YaCy or Faroo [86, 87] also distribute search engine components among participating peers. YaCy implements decentralized crawling and indexing by the use of a local web proxy, which records outgoing HTTP requests, crawls the resources, and stores the information in a database serving as a local index. The local indices are being distributed among peers to ensure reliability by redundancy. Their entirety forms the global index, which can be searched by respective P2P protocols. YaCy's search and ranking methods are mainly based on keyword metrics, such as term appearances in titles, URLs, hyperlink anchor texts, and simple statistics, such as number and ratio of keywords in a document, number of stop words, and number of in and outbound links. Similar to other bag-of-words methods the search performance with regard to user intent, disambiguation, similarity, or lingo is questionable. Furthermore, [87] showed that YaCy, and other distributed web search engines are prone to censorship and have privacy vulnerabilities. Countermeasures, as the *node density protocol* and the *webpage verification protocol* to detect malicious nodes and censorship on node-level, were proposed.

1.6 Conclusion

This paper presented an overview of relevant aspects, that need to be considered when designing a new, state-of-the-art web search engine or recommender system. One of the main aspects that were addressed is a taxonomy of web search, consisting of user intention, search subject, and use-cases. This was conducted under consideration of the structure of the web, which, as turned out, is a complex information space composed of unstructured data containing relevant and irrelevant information stored in accessible and hidden repositories. Furthermore, fundamental functional building blocks, crucial to supporting web search were presented. Subsequently, the focus was put on search engine algorithms and implementations, discussing methods as, how to select a suitable result set from an index, and how to derive relevance scores for ranking. Lastly, approaches to distribute search in a decentralized P2P network were presented. Figure 1.4 illustrates the resulting overarching classification of the web derived from the aforementioned aspects and puts them into relation.

Further research in the field of decentralized search is not only motivated by concerns regarding transparency and control of dominating centralized search providers, but also by the continuously altering and growing structure of the web, which makes it increasingly difficult and costly to maintain a comprehensive index. It was shown, that especially clustering and linking of related resources on the web, is crucial for search, i.e., identifying a suitable set of search results. The approach presented in TheBrain achieves this by utilizing a distributed graph database. Since it relies on an active P2P network, it brings concerns with respect to implementation effort, availability, computational cost, and similar. Therefore, more static approaches, like the WebMap, are needed, to achieve a distributed, and uniform linking and clustering scheme for related resources. The next steps will be to continue work in this context, i.e., implementing a proof of concept for the WebMap, and develop efficient search and routing algorithms.

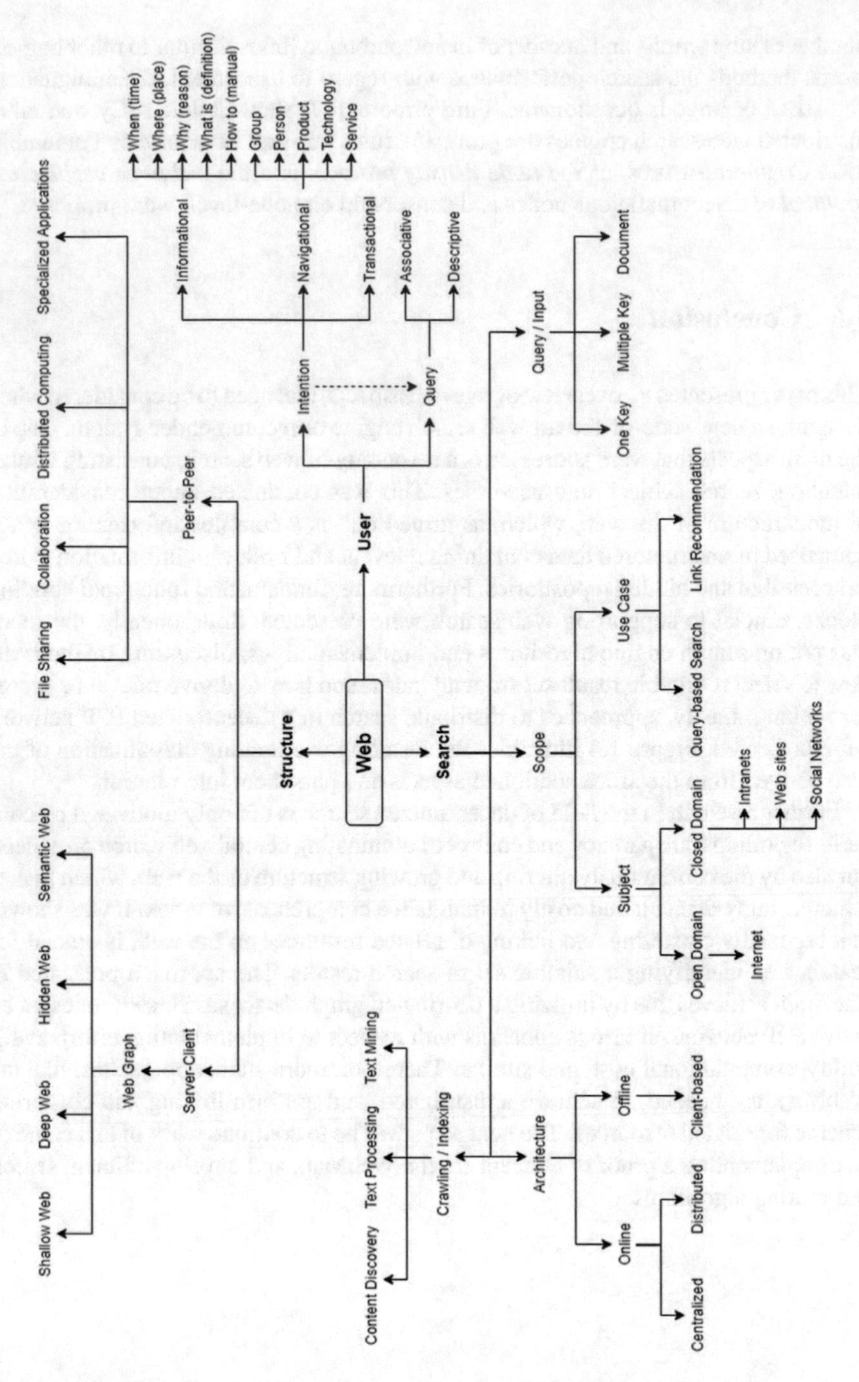

Fig. 1.4 Classification of the web with respect to structure and search

References

1. Roßrucker, G.P., Unger, H.: Webmap: A concept for webengine version 3.0. In: The Autonomous Web (2021)
2. Berners-Lee, T., Cailliau, R., Groff, J.-F., Pollermann, B.: World-wide web: The information universe. Electron. Netw. Res. Appl. Policy **8**(2) (1992). Westport, Spring
3. Matthew Gray.: Web growth summary (1996). http://www.mit.edu/people/mkgray/net/web-growth-summary.html. Accessed 03 Feb 2021
4. Internet Live Stats.: Total number of websites. https://www.internetlivestats.com/total-number-of-websites/. Accessed 03 Feb 2021
5. Van den Bosch, A., Bogers, T., De Kunder, M.: Estimating search engine index size variability: a 9-year longitudinal study. Scientometrics **107**(2), 839–856 (2016)
6. Broder, A., Kumar, R., Maghoul, F., Raghavan, P., Rajagopalan, S., Stata, R., Tomkins, A., Wiener, J.: Graph structure in the web. Comput. Netw. **33**(1), 309–320 (2000)
7. Meusel, R., Vigna, S., Lehmberg, O., Bizer, C.: Graph structure in the web—revisited: A trick of the heavy tail. In: Proceedings of the 23rd International Conference on World Wide Web, WWW '14 Companion, pp. 427–432. New York, NY, USA (2014). Association for Computing Machinery
8. Meusel, R., Vigna, S., Lehmberg, O., Bizer, C.: The graph structure in the web—analyzed on different aggregation levels. J. Web Sci. **1** (2015)
9. Bergman, M.K.: The Deep Web: Surfacing Hidden Value (2000)
10. Berners-Lee, T., Hendler, J., Lassila, O.: The semantic web. Sci. Am. **284**(5), 34–43 (2001)
11. Pourebrahimi, B., Bertels, K., Vassiliadis, S.: A survey of peer-to-peer networks. In: Proceedings of the 16th Annual Workshop on Circuits, Systems and Signal Proessing (2005)
12. Ripeanu, M., Foster, I., Iamnitchi, A.: Mapping the gnutella network: Properties of large-scale peer-to-peer systems and implications for system design. IEEE Internet Comput. J. **6**, 10 (2002)
13. Saroiu, S., Gummadi, K.P., Gribble, S.D.: Measuring and analyzing the characteristics of napster and gnutella hosts. Multimed. Syst. **9**(2), 170–184 (2003)
14. Cohen, B.: Bittorrent-a new p2p app. Yahoo eGroups (2001)
15. Lueg, C., Fisher, D.: From Usenet to CoWebs: Interacting with Social Information Spaces. Springer Science & Business Media (2012)
16. Clarke, I., Sandberg, O., Wiley, B., Hong, T.W.: Freenet: A distributed anonymous information storage and retrieval system. In: Designing Privacy Enhancing Technologies, pp. 46–66. Springer, Berlin (2001)
17. Abdul-Rahman, A.: The pgp trust model. EDI-Forum: J. Electron. Commerce **10**, 27–31 (1997)
18. Nakamoto, S., Bitcoin, A.: A peer-to-peer electronic cash system. Bitcoin, vol. 4 (2008). https://bitcoin.org/bitcoin.pdf
19. Vahdat, A., Anderson, T., Dahlin, M., Belani, E., Culler, D., Eastham, P., Yoshikawa, C.: Webos: Operating system services for wide area applications, pp. 52 – 63, 08 (1998)
20. Kropf, P., Unger, H., Babin, G.: Wos: an internet computing environment. In: Workshop on Ubiquitous Computing, IEEE International Conference on Parallel Architecture and Compilation Techniques, pp. 1422–1430. Institute of Electrical and Electronics Engineers (IEEE) (2000)
21. Kropf, P., Plaice, J., Unger, H.: Towards a Web Operating System (wos). 02 (2002)
22. Nelson, B.J.: Remote Procedure Call. Carnegie Mellon University (1981)
23. Plášil, F., Stal, M.: An architectural view of distributed objects and components in corba, java rmi and com/dcom. Softw Concepts Tools **19**(1), 14–28 (1998)
24. Xu, G., Cockburn, A., Mckenzie, B.: Lost on the Web: An Introduction to Web Navigation Research (2001)
25. Broder, A.: A taxonomy of web search. SIGIR Forum **36**, 3–10 (2002)
26. Jansen, B.J., Booth, D.L., Spink, A.: Determining the user intent of web search engine queries. In: Proceedings of the 16th International Conference on World Wide Web, WWW '07, pp. 1149–1150. New York, NY, USA (2007). Association for Computing Machinery

27. Li, H., Cao, Y., Xu, J., Hu, Y., Li, S., Meyerzon, D.: A new approach to intranet search based on information extraction. In: CIKM '05 (2005)
28. Baeza-Yates, R., Calderón-Benavides, L., González-Caro, C.: The intention behind web queries. In: International Symposium on String Processing and Information Retrieval, pp. 98–109. Springer, Berlin (2006)
29. Kleinberg, J.M.: Authoritative sources in a hyperlinked environment. J. ACM **46**(5), 604–632 (1999)
30. Page, L., Brin, S., Motwani, R., Winograd, T.: The pagerank citation ranking: Bringing order to the web. Technical Report 1999-66, Stanford InfoLab, November (1999). Previous number = SIDL-WP-1999-0120
31. Firoozeh, N., Nazarenko, A., Alizon, F., Daille, B.: Keyword extraction: Issues and methods. Nat. Lang. Eng. **26**(3), 259–291 (2020)
32. Hawking, D.: Challenges in Enterprise Search (2004)
33. Stenmark, D.: Method for intranet Search Engine Evaluations. pp. 7–10 (1999)
34. Zwol R.V., Apers, P.: Searching documents on the intranet. In: WOWS (1999)
35. Hedström, U., Borin, L., Olsson, F.: Automatic Generation of Metadata for Indexing, Searching and Navigating in an Intranet (2001)
36. Zhu, H., Raghavan, S., Vaithyanathan, S., Löser, A.: Navigating the intranet with high precision. In: Proceedings of the 16th International Conference on World Wide Web, WWW '07, pp. 491–500. New York, NY, USA (2007). Association for Computing Machinery
37. Dignum, S., Kruschwitz, U., Fasli, M., Kim, Y., Song, D., Beresi, U.C., de Roeck, A.: Incorporating seasonality into search suggestions derived from intranet query logs. In: 2010 IEEE/WIC/ACM International Conference on Web Intelligence and Intelligent Agent Technology, vol. 1, pp. 425–430 (2010)
38. Xue, G.-R., Zeng, H.-J., Chen, Z., Ma, W.-J., Zhang, H.-J., Lu, C.-J.: Implicit link analysis for small web search. In: Proceedings of the 26th Annual International ACM SIGIR Conference on Research and Development in Informaion Retrieval, SIGIR '03, pp. 56–63. New York, NY, USA (2003). Association for Computing Machinery
39. Brenes, D.J., Gayo-Avello, D., Pérez-González, K.: Survey and evaluation of query intent detection methods. In: Proceedings of the 2009 Workshop on Web Search Click Data, WSCD '09, pp. 1–7. New York, NY, USA (2009). Association for Computing Machinery
40. Hafernik, C.T., Jansen, B.J.: Understanding the specificity of web search queries. In: CHI '13 Extended Abstracts on Human Factors in Computing Systems, CHI EA '13, pp. 1827–1832, New York, NY, USA (2013). Association for Computing Machinery
41. Roy, R.S., Choudhury, M., Bali, K.: Are web search queries an evolving protolanguage? In: The Evolution Of Language, pp. 304–311. World Scientific (2012)
42. Pazzani, M.J., Billsus, D.: Content-based recommendation systems. In: The Adaptive Web, pp. 325–341. Springer, Berlin (2007)
43. Georg, P.: Roßrucker. Towards a new link recommendation indicator. The Autonomous Word Wide Web (2021)
44. Sivapalan, S., Sadeghian, A., Rahnama, H., Madni, A.M.: Recommender systems in e-commerce. In: 2014 World Automation Congress (WAC), pp. 179–184 (2014)
45. Sarwar, B., Karypis, G., Konstan, J., Riedl, J.: Analysis of recommendation algorithms for e-commerce. In: Proceedings of the 2nd ACM Conference on Electronic Commerce, EC '00, pp. 158–167. New York, NY, USA (2000). Association for Computing Machinery
46. Herlocker, J.L., Konstan, J.A., Terveen, L.G., Riedl, J.T.: Evaluating collaborative filtering recommender systems. ACM Trans. Inf. Syst. **22**(1), 5–53 (2004)
47. Das, A.S., Datar, M., Garg, A., Rajaram, S.: Google news personalization: Scalable online collaborative filtering. In: Proceedings of the 16th International Conference on World Wide Web, WWW '07, pp. 271–280. New York, NY, USA (2007). Association for Computing Machinery
48. Cho, J.: Crawling the Web: Discovery and Maintenance of Large-Scale Web Data. Computer Science. Stanford University (2001)

49. Pant, G., Srinivasan, P., Menczer, F.: Crawling the web. In: Web Dynamics, pp. 153–177. Springer, Berlin (2004)
50. Hernández, I., Rivero, C.R., Ruiz, D.: Deep web crawling: a survey. World Wide Web **22**(4), 1577–1610 (2019)
51. Raghavan, S., Garcia-Molina, H.: Crawling the hidden web. In: 27th International Conference on Very Large Data Bases (VLDB 2001) (2001)
52. Srividhya, V., Anitha, R.: Evaluating preprocessing techniques in text categorization. Int. J. Comput. Sci. Appl. **47**(11), 49–51 (2010)
53. Vijayarani, S., Ilamathi, M.J., Nithya, M.: Preprocessing techniques for text mining-an overview. Int. J. Comput. Sci. Commun. Netw. **5**(1), 7–16 (2015)
54. Ladani, D.J., Desai, N.P.: Stopword identification and removal techniques on tc and ir applications: A survey. In: 2020 6th International Conference on Advanced Computing and Communication Systems (ICACCS), pp. 466–472 (2020)
55. Moral, C., de Antonio, A., Imbert, R., Ramírez, J.: A survey of stemming algorithms in information retrieval. Inf. Res. Int. Electron. J. **19**(1) (2014)
56. Rajaraman, A., Ullman, J.D.: Mining of Massive Datasets. Cambridge University Press, Cambridge (2011)
57. Jin, W., Srihari, R.K.: Graph-based text representation and knowledge discovery. In: Proceedings of the 2007 ACM Symposium on Applied Computing (2007)
58. Kubek, M., Unger, H.: Centroid terms as text representatives. In: Proceedings of the 2016 ACM Symposium on Document Engineering, pp. 99–102 (2016)
59. Mihalcea, R., Tarau, P.: Textrank: Bringing order into text. In: Proceedings of the 2004 Conference on Empirical Methods in Natural Language Processing, pp. 404–411 (2004)
60. Slobodan, B., Meštrović, A., Sanda, M.: An overview of graph-based keyword extraction methods and approaches. J. Inf. Org. Sci. **39**, 1–20 (2002)
61. Simcharoen, S., Ruamsuk, Y., Mingkhwan, A., Unger, H.: Modeling a hierarchical abstraction process on top of co-occurrence graphs. In: 2019 Research, Invention, and Innovation Congress (RI2C), pp. 1–5 (2019)
62. Schaeffer, S.E.: Graph clustering. Comput. Sci. Rev. **1**(1), 27–64 (2007)
63. Koutra, D., Parikh, A., Ramdas, A., Xiang, J.: Algorithms for graph similarity and subgraph matching. Proc. Ecol. Inference Conf. **17** (2011)
64. Le, Q., Mikolov, T.: Distributed representations of sentences and documents. In: International Conference on Machine Learning, pp. 1188–1196. PMLR (2014)
65. Mikolov, T., Chen, K., Corrado, G., Dean, J.: Efficient estimation of word representations in vector space (2013). ArXiv:1301.3781
66. Kubek, M.M., Böhme, T., Unger, H.: Spreading activation: a fast calculation method for text centroids. In: Proceedings of the 3rd International Conference on Communication and Information Processing, pp. 24–27 (2017)
67. Tang, H., Tan, S., Cheng, X.: A survey on sentiment detection of reviews. Exp. Syst. Appl. **36**(7), 10760–10773 (2009)
68. Pang, B., Lee, L., Vaithyanathan, S.: Thumbs Up? Sentiment Classification Using Machine Learning Techniques (2002). ArXiv:cs/0205070
69. Nasukawa, T., Yi, J.: Sentiment analysis: Capturing favorability using natural language processing. In: Proceedings of the 2nd International Conference on Knowledge Capture, K-CAP '03, pp. 70–77. New York, NY, USA (2003). Association for Computing Machinery
70. Kennedy, A., Inkpen, D.: Sentiment classification of movie reviews using contextual valence shifters. Comput. Intell. **22**(2), 110–125 (2006)
71. De Bra, P.M.E., Post, R.D.J.: Searching for Arbitrary Information in the www: The Fish-Search for Mosaic (1994)
72. Hersovici, M., Jacovi, M., Maarek, Y.S., Pelleg, D., Shtalhaim, M., Ur, S.: The shark-search algorithm. an application: tailored web site mapping. Comput. Netw. ISDN Syst. **30**(1), 317–326 (1998). Proceedings of the Seventh International World Wide Web Conference
73. Brin, S., Page, L.: The anatomy of a large-scale hypertextual web search engine. Comput. Netw. ISDN Syst. 30–1 (1998)

74. Qin, T., Liu, T.-Y., Zhang, X.-D., Chen, Z., Ma, W.-Y.: A study of relevance propagation for web search. In: Proceedings of the 28th Annual International ACM SIGIR Conference on Research and Development in Information Retrieval, SIGIR '05, pp. 408–415. New York, NY, USA (2005). Association for Computing Machinery
75. Shakery, A., Zhai, C.: Relevance propagation for topic distillation uiuc trec 2003 web track experiments. In: TREC, pp. 673–677. Citeseer (2003)
76. Song, R., Wen, J.-R., Shi, S., Xin, G., Liu, T,-Y.: Microsoft research asia at web track and terabyte track of trec 2004. In: 2004 Text REtrieval Conference (TREC'04) (2004)
77. Smyth, B., Balfe, E., Freyne, J., Briggs, P., Coyle, M., Boydell, O.: Exploiting query repetition and regularity in an adaptive community-based web search engine. User Model User-Adapt Interaction **14**(5), 383–423 (2004)
78. Dupret, G., Liao, C.: A model to estimate intrinsic document relevance from the clickthrough logs of a web search engine. In: Proceedings of the Third ACM International Conference on Web Search and Data Mining, WSDM '10, pp. 181–190. New York, NY, USA (2010). Association for Computing Machinery
79. Vassilvitskii, S., Brill, E.: Using web-graph distance for relevance feedback in web search. In: Proceedings of the 29th Annual International ACM SIGIR Conference on Research and Development in Information Retrieval, SIGIR '06, pp. 147–153. New York, NY, USA (2006). Association for Computing Machinery
80. Kubek, M., Böhme, T., Unger, H.: Empiric experiments with text representing centroids. Lecture Notes on Information Theory **5**(1) (2017)
81. Takaki, T., Fujii, A., Ishikawa, T.: Associative document retrieval by query subtopic analysis and its application to invalidity patent search. In: Proceedings of the thirteenth ACM international conference on Information and knowledge management, pp. 399–405 (2004)
82. Dean, J., Ghemawat, S.: Mapreduce: simplified data processing on large clusters. Commun. ACM **51**(1), 107–113 (2008)
83. Stoica, I., Morris, R., Karger, D., Kaashoek, M.F., Balakrishnan, H.: Chord: A scalable peer-to-peer lookup service for internet applications. ACM SIGCOMM Comput. Commun. Rev. **31**(4), 149–160 (2001)
84. Kubek, M.: Concepts and Methods for a Librarian of the Web. Springer, Berlin (2019)
85. Simcharoen, S., Unger, H.: The Brain: WebEngine Version 2.0. In: The Autonomous Web (2021)
86. Herrmann, M., Ning, K.-C., Diaz, C., Preneel, B.: Description of the yacy distributed web search engine. Technical Report, KU Leuven ESAT/COSIC, IMinds (2014)
87. Herrmann, M., Zhang, R., Ning, K.-C., Diaz, C., Preneel, B.: Censorship-resistant and privacy-preserving distributed web search. In: 14th IEEE International Conference on Peer-to-Peer Computing, pp. 1–10. IEEE (2014)

Chapter 2
A Concept for Recommender Systems Based on Text-Representing Centroids

Herwig Unger, Mario Kubek, Yanakorn Ruamsuk, and Anirach Mingkhwan

2.1 Motivation and Background

Fast developments in science and technology make decision-making a more and more complex, tedious task, mainly since the permanently growing number of influencing parameters and their manifold interdependencies are hard to overseen and understand. In particular young, untrained or non-specialised people have tremendous problems meeting the right decision in the available time, usually much too short of obtaining and reading all available and/or existing documents on respective objects and processes. While in some cases a wrong decision can be neglected or corrected, in other cases (e.g. in medicine), it may cause danger for the health and life of people.

Recommender systems [1] are one possibility to support those complex decision processes of humans by information technology. However, building feasible software requires beside excellent programmers a significant apriori knowledge of experts, a large set of learning samples including a learning process by a long term observation of the objects and processes they are involved.

While the first generation of recommender system required the coding of expert knowledge within programming code or rules (including fuzzy rules [2]) and were difficult to update by daily new knowledge, modern Ai- and often neural network and the deep learning-based system can learn by themself in a supervised or unsupervised manner. Nevertheless, those systems usually fail to explain why a decision was made, which factors have in which manner influenced the decision and do not allow to integrate the recent user's experiences to the knowledge base. As a result, trustworthy and explainable AI [3, 4] aim to make the decision process more transparent and allow the user to trace and reproduce each step of the a decision like a process of logical argumentation.

H. Unger and M. Kubek, *The Autonomous Web*, Studies in Big Data 101,
https://doi.org/10.1007/978-3-030-90936-9_2

The following four criteria reflect the needs for the design of modern recommender systems:

1. The system shall be able to process knowledge presented in natural language documents.
2. A permanent background knowledge processing and learning process with a low grade of supervision shall be supported.
3. Faked expired or replaced information shall be automatically removed or replaced in the system.
4. Decision making shall be made understandable in an interactive process, allowing a human-centred update or change of the contained knowledge.

In this paper, a new, brain-inspired method of natural language processing derived from the well-known modular segregation, which can also be observed in co-occurrence graphs: words belonging to the same problem usually appear at a short distance from each other and build clusters with respective cluster centres, while the meaning of any sentence or query may be assigned to those clusters and/or nodes by using the so-called text-represented centroids, a particular form of so-called generalised medians.

Since co-occurrence graphs can be extended by reading different sources in the continuously in the background, they are the perfect knowledge base for a new class of recommender systems belonging to trustworthy and explainable AI.

2.2 Fundamentals of Natural Language Processing

2.2.1 Text-Representing Centroids

In [5–8] the definition, properties, use and fast calculation text-representing centroids (TRC) have been exhaustingly discussed, such that a citation from those publications shall summarise the most important definitions and findings in this section.

«Any two words w_i and w_j are called co-occurrent, if they appear in one sentence (or any other well defined environment or context) together. This co-occurrence relation may be used to define a graph $G = (W, E)$. Therefore, the set words of a documents corresponds to the set of nodes $w_a \in W$ and two nodes are connected by an edge $(w_a, w_b) \in E$, iff w_a and w_b are co-occurrent. A weight function $g((w_a, w_b))$ can be introduced to represent the frequency of a co-occurrence in a document, while usually only co-occurrences of a high significance $\sigma > 1, \sigma \leq g((w_a, w_b))$ are taken into account.

In a next step, a *distance* must be defined in G. Two words are close, if $g((w_a, w_b))$ is high. If $(w_a, w_b) \in E$ (i.e. the words co-occurrent) their distance $d(w_a, w_b)$ is easily to be defined as

$$d(w_a, w_b) = \frac{1}{g((w_a, w_b))}.$$

Otherwise, let us consider the shortest path $p = \{(w_1, w_2), (w_2, w_3), \ldots, (w_k, w_{k+1})\}$ with $w_1 = w_a, w_{k+1} = w_b$ and $(w_i, w_{i+1}) \in E$ and define

$$d(w_a, w_b) = \sum_{i=1}^{k} d((w_i, w_{i+1}))$$

The[1] definition of the centroid term $\chi(D)$ of a document D uses all N words $w_1, w_2, \ldots, w_N \in D$, which can be reached from any term $t \in G$. Therefore, the average distance of all words of D to the term t, $d(D, t)$, can be obtained by

$$d(D, t) = \frac{\sum_{i=1}^{N} d(w_i, t)}{N},$$

The centroid term $\chi(D)$ is defined to be the term with

$$d(D, \chi(D)) = MINIMAL.$$

Note, that not necessarily $\chi \in D$.

Let χ_1 be the centroid term of D_1, and χ_2 the centroid term of D_2, then $d(\chi 1, \chi 2)$ can be understood as the distance of the two documents D_1 and D_2.»

2.2.2 Inference Mechanism

For the use in recommender systems, it may make sense to label some words especially as follows:

- a set $I \subset W$ of significant input keywords in order to describe the decision problem formulated by the user. In the case of a medical recommender system this would be the symptoms of the disease and
- a set $O \subset W$ of output words which belong to descriptive words of the target/recommendation, again in case of a medical problem or recommender system the name of possible, diagnosable diseases and/or treatments.

Initial sets of words for I and O can easily be defined by experts or obtained from web pages of companies or associations (like in our example, the lists of symptoms and diseases from the Centers for Disease Control and Prevention(CDC) in Atlanta. At the same time, also parameter lists of technical devices and existing products might be feasible). Also, those terms may be learnt or extended from specially designed systems as described in [9, 10].

The decision making (inference) mechanism can be described now easily by the following steps:

[1] If there is no path between any two words w_a and w_b, $d(w_a, w_b) = \infty$ shall be set.

1. Ask the user for descriptive input on the current decision situation and support his decision with close words $w \in I$.
2. Built the TRC of the input information.
3. Find target words (objects) in the co-occurrence graph close to the found TRC of the input information.
4. Present a ranking of possible output terms $w \in O$ depending on their distance from the input's TRC.
5. If the user is not satisfied, suggest more related input criteria from words $w \in I$, which are nearby/neighbours in the co-occurrence graph and also add respective attributes, which could be parametrised and go to 1.
6. END.

Of course, the mechanism above gives in step 5 the opportunities for manifold user interactions. Once a target objects $o \in O$ is prioritised, similar, not yet considered input words may be obtained from its environment in the co-occurrence graph, possibly with some needed, more descriptive attributes (e.g. high fever, fever above 40 degrees etc.). Also, similar, i.e. close, other target objects can be considered and result in differential analysis of the decision situation.

Last but not least shall be mentioned that a fast executable update of the co-occurrence graph by new words and co-occurrences extracted from newly processed documents in the background will result in an immediate update of the knowledge of the system and possibly qualitative better answers.

2.3 Design Example: The Medical Recommender System

2.3.1 The Medical Dilemma

For some reasons the authors were motivated to apply the new recommender approach at first for the diagnosis of diseases. The so called medical dilemma is known to most people: young doctors, which just had finished their studies, find themself immediately in the daily life of big hospitals or a doctors practice. Still inexperienced, they are confronted with a manifold of symptoms, which can often not clearly and explicitly assigned to a given disease as in textbook examples or demonstrations. Moreover, some diseases may have very similar symptoms, which are hard to distinguish or related to seldom diseases, a young medical doctor may not know. In such a manner, the risk of mistakes and wrong treatments is significantly increased. Related to that problem, people usually prefer older, experienced physicians.

As it is shown in Fig. 2.1 they have trained their brain after the initial study by a lot of seen cases, different appearances of those diseases and also misleading symptoms. Sometimes, try and fail is still an important learning strategy in medicine handling a complex system like the human body.

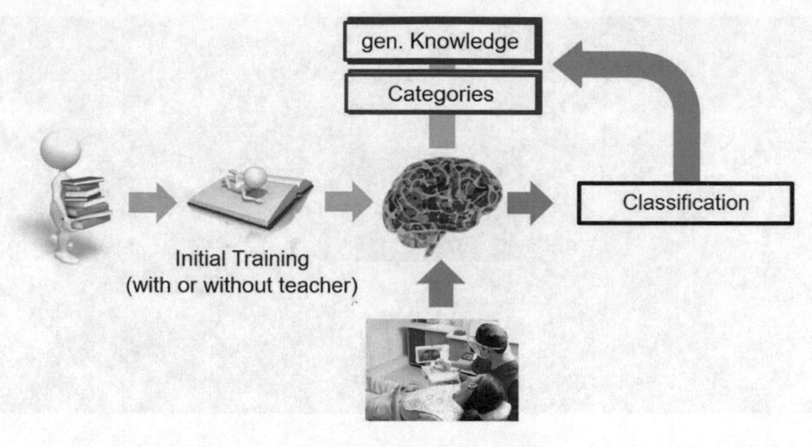

Fig. 2.1 Learning and knowledge management of the brain

2.3.2 *Structure of the Medical Recommender System*

Standard recommender systems today usually consist of three components:

- a database containing the knowledge of the system in any special, coded form, e.g. rules, fuzzy rules, weights of a neural network
- an interface realising the action with the user including the handling of the user queries, the presentation of a result in a subject-specific presentation, which can be understood by the respective experts and
- an inference machine, which maps the user's query to a respective answer.

Note that in the standard case of a recommender system, no learning from recently processed queries or the experience of the current user as shown in Fig. 2.1 is possible.

Consequently, the design of the new recommender system follows that major design with some changes and updates for the data storage and a newly added online learning component:

1. The data in the new class of recommender systems is a co-occurrence graph built from literature, documentation and other written text sources related to the subject of the recommender system.
   Newly developed graph databases, like Neo4J [11] or NetworkX [12] are very suitable to handle those large graphs with up to 500,000 nodes and the respective numbers of edges and also offer standard operations on it, e.g. finding the shortest path.
2. The inference system performs the matching between queries of the user and the wanted output by determining TRC's and a consideration of distances in the recent co-occurrence graph.
3. A suitable user interface as it can be seen in Fig. 2.2. Differing from a classical recommender system, this interface is a significant change since it tries to represent

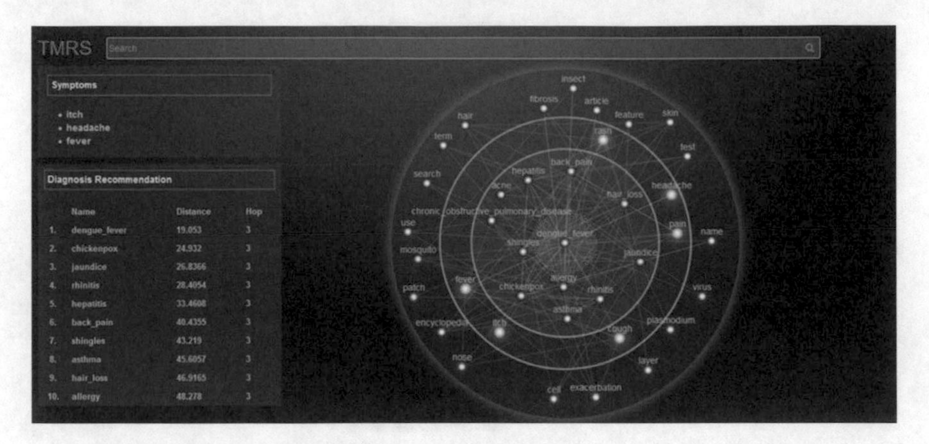

Fig. 2.2 User interface of the medical recommender system

and explain the connections between the input (symptoms, green dots in the figure and upper table on the left side of the interface window) and the results (diseases, red dots in the figure). Herby, the most probable diagnosis is centred and the red circled in the middle, while other, also possible, similar and therefore related results (diseases with similar symptoms) are shown around. Finally, other related, more general, but frequently used terms (blue dots) in the context of the respective topic are given in the outer shell. In such a manner, concentric circles are used to separate the different terms while still trying to keep the right distance within each shell. The colour of the edges shows once more the intensity (distance of nodes) of each connection by different colours using grey for less, distant nodes and changing to red for very strongly related, close nodes.

User interactions are possible since not only the given symptoms are presented, but any other related to the diagnosed disease are shown as well as related other diseases. by clicking on the nodes of symptoms, more attributes (e.g. high fever, fever above 40 degrees, low temperature) can be specified to refine the diagnosis. By clicking on a diseases' node, some diagnoses may be excluded (to reduce interesting symptoms) and a a detailed description of the disease appears on the screen.

Last but not least, the system uses distances between nodes—and in particular distances to the centred, main diagnosed disease (also in the table on the left side of the screen)—to differ the diagnosis probability from other ones.

4. Since the described system can process any related textual data by extracting the respective (new) words and co-occurrences from the document, an easy online and background learning process will be possible.

   Therefore it is only necessary to equip the system with an Internet access as well as a target-oriented, standard web-crawler, which retrieves and process new documents from related sources. This process can happen in the background, while the update of the knowledge base (i.e. the co-occurrence graph) can be done in a very short time while the system is used simultaneously. Oblivion based mechanisms as described in [] may ensure that wrong, outdated or faked news are automatically neglected and/or removed from the knowledge base.

It is easy to be seen that with the above-made design also other target areas like product recommender systems or mistake finding routines, e.g. to solve car problems can be designed.

2.3.3 *First Experiences*

First empiric results show already the success of the described approach. Therefore 225 typical, often appearing diseases were considered. The respective descriptions with symptoms, diagnosis, treatments and additional information (like epidemiology, prevention) were just downloaded from Wikipedia and processed a priori in an almost unchanged manner directly in the knowledge basis, i.e. the co-occurrence graph of the system. Symptoms and disease were tagged using the corresponding lists of the CDC, while diseases were only considered, if enough knowledge (i.e. an article with the respective title) was processed. This corresponds to the experience of the authors that without a minimal size of the co-occurrence graph (approximately 100–200 documents need to be processed) and a sufficient knowledge on every disease (and therefore a right embedding and stability of the edge weights in the co-occurrence graph) no stable determinations of the TRC's are possible.

In a second step, several sets of symptoms (usually three) were presented to the system. In Fig. 2.3 the obtained results are shown for the first four, most probable

Symptoms	1st Candidate	2nd Candidate	3rd Candidate	4th Candidate
itch, headache, fever	Dengue fever (19.05)	Chickenpox (24.93)	Jaundice (26.83)	Rhinitis (28.40)
cough, headache, fever	Influenza (8.62)	Common cold (13.95)	Pneumonia (14.68)	Sinusitis (17.54)
stomach, pain, vomit	Peptic ulcer disease (7.10)	Gastroenteritis (19.51)	Pancreatitis (21.07)	Influenza (24.40)
itch, redness, foot	Athlete's foot (5.99)	Dermatophytosis (15.70)	Scabies (24.79)	Dermatitis (33.73)
headache, memory, loss	Transient ischemic attack (14.88)	Traumatic brain injury (15.42)	Stroke (23.68)	Encephalitis (31.10)
difficulty, walk, slowness, movement	Parkinson's disease (13.72)	Dementia (27.76)	Osteoarthritis (29.39)	Sleep disorder (32.99)
fear, water, confusion	Rabies (10.09)	Meningitis (35.95)	Anemia (36.56)	Sleep paralysis (40.30)
pain, breathlessness, cough	Pneumonia (24.80)	Heart failure (30.99)	Sinusitis (39.38)	Asthma (41.71)
weight, loss, weakness	Amyotrophic lateral sclerosis (22.63)	Stroke (30.78)	Anemia (31.16)	Atherosclerosis (34.13)
anxiety, mania, paranoia	Schizophrenia (24.82)	Major depressive disorder (33.46)	Bipolar disorder (42.70)	Insomnia (44.35)

Fig. 2.3 Examples for a Symptom-Diagnosis Output

results, while in brackets the distance of the TRC determined from the input data to the corresponding diseases' node is presented. As it is clearly to be seen even to the medical layman, those results definitely fit the assumed symptoms. Also, a significant distance between the major (first) candidate and the other possible diagnosis is to be seen, i.e. a nice separation between the diagnoses exists. It is also important that together with more or less simple, harmless diseases also dangerous, life-threatening (twin-) diagnoses are shown. The obtained results were stable with the basic implementation of the system. Even if updates in the co-occurrence graph has been made, the major diagnosis remains unchanged. Also, definite wrong diagnoses were never obtained.

Further improvements might include the probability of a diagnosed disease in the result generation and also prioritise hints to life-threatening diseases with similar results. However, in order to demonstrate the powerfulness of the described method, such considerations were not implemented yet.

2.4 Conclusion

An innovative design concept for a medical recommender system based on the usage of so-called text-representing centroids has been presented. Differing from other recommender systems, it is able to gain knowledge from any textual information and does not need experts and the transformation of expert knowledge into computer-processable formats. Also, a direct background learning process without offline learning periods is integrated. As the included decision making process relies on a distance measure applied on a (large) graph, its outputs are both trustworthy and explainable, an important aspect in the area of explainable artificial intelligence. With its goal to support medical diagnoses, the first prototype of this new recommender system already achieved high accuracy and stability. Further applications in the area of product marketing, system diagnostics and configuration seem to be immediately possible and realisable.

References

1. Ricci, F., Rokach, L., Shapira, B.: Introduction to Recommender Systems Handbook, pp. 1–35. Springer, Berlin (2011). Recommender Systems Handbook
2. Mendel, J.M.: Uncertain Rule-based Fuzzy Logic Systems : Introduction and New Directions. Prentice Hall PTR. ISBN 978-0130409690
3. Russell S.J., Norvig P.: Artificial Intelligence: A Modern Approach, 2nd edn. Prentice Hall. ISBN 0-13-790395-2
4. Loyola-González, R.O.: Black-Box versus white-box: understanding their advantages and weaknesses from a practical point of view. In: IEEE Access, vol. 7, pp. 154096–154113 (2019). https://doi.org/10.1109/ACCESS.2019.2949286

5. Kubek, M.M., Unger, H.: Centroid terms as text representatives. In: Proceedings of the 2016 ACM Symposium on Document Engineering, DocEng '16, pp. 99–102. ACM, New York, NY, USA (2016)
6. Kubek, M.M., Unger, H.: Centroid terms and their use in natural language processing. In: Autonomous Systems 2016, Fortschritt-Berichte VDI, Reihe 10 Nr. 848, pp. 167–185, VDI-Verlag Düsseldorf (2016)
7. Kubek, M.M., Böhme, T., Unger, H.: Empiric experiments with text representing centroids. Lect. Notes Inf. Theory **5**(1), 23–28 (2017). https://doi.org/10.18178/lnit.5.1.23-28
8. Kubek, M.M., Böhme, T., Unger, H.: Spreading activation: a fast calculation method for text centroids. In: Proceedings of the 3rd International Conference on Communication and Information Processing (ICCIP 2017). ACM, New York, NY, USA (2017)
9. Teng, F., Ma, M., Ma, Z., Huang, L., Xiao, M., Li, X.: A Text annotation tool with pre-annotation based on deep learning. In: International Conference on Knowledge Science, Engineering and Management, pp. 440–451. Springer, Cham (2019)
10. Zhou, B., Cai, X., Zhang, Y., Guo, W., Yuan, X.: newblock MTAAL: Multi-task adversarial active learning for medical named entity recognition and normalization. Proc. AAAI Conf. Artif. Intell. **35**(16), 14586–14593 (2021)
11. NEO4J,GRAPH DATA PLATFORM, Accessed 20 July 2021. https://neo4j.com/
12. NetworkX,Network Anylysis in Python, Accessed 20 July 2021. https://networkx.org/

Chapter 3
WebEngine Version 1.0: Building a Decentralised Web Search Engine

Mario Kubek and Herwig Unger

3.1 Introduction

It is definitely a great merit of the World Wide Web (WWW, web) to make the world's largest collection of documents of any kind in digital form easily available at any time and any place without respect to the number of copies needed. It can therefore be considered to be the knowledge base or library of mankind in the age of information technology. Google (https://www.google.com/), as the world's largest and most popular web search engine with its main role to connect information and the place/address where it can be found, might be the most effective, currently available information manager.

Even so, in the authors' opinion, Google and Co. are just the mechanistic, brute force answer to the problem of effectively managing the complexity of the WWW and handling its big data volumes. As already discussed e.g. in [1], a copy of the web is established by crawling it and indexing web content in big reverse index files containing for each occurring word a list of files in which they appear. Complex algorithms try to find those documents that contain all words of a given query and closely related ones. Since (simply chosen) keywords/query terms appear in millions of (potentially) matching documents, a relevance ranking mechanism must avoid that all of these documents are touched and presented in advance to the user (see Fig. 3.1).

In the ranking process, the content quality and relative position of a document in the web graph as well as the graph's linking structure are taken into account as important factors. In addition to organic search results obtained from this process, advertisements are often presented next to them which are related to the current query or are derived from personalisation efforts and detected user's interests. In both cases, web search engines do not take into account probably existing (local) user knowledge. To a certain extent, this procedure follows a top-down approach as this filtering is applied on the complete index for each incoming query in order to return a ranked list of links to matching documents. The top-ranked documents in this list are generally useful.

H. Unger and M. Kubek, *The Autonomous Web*, Studies in Big Data 101,
https://doi.org/10.1007/978-3-030-90936-9_3

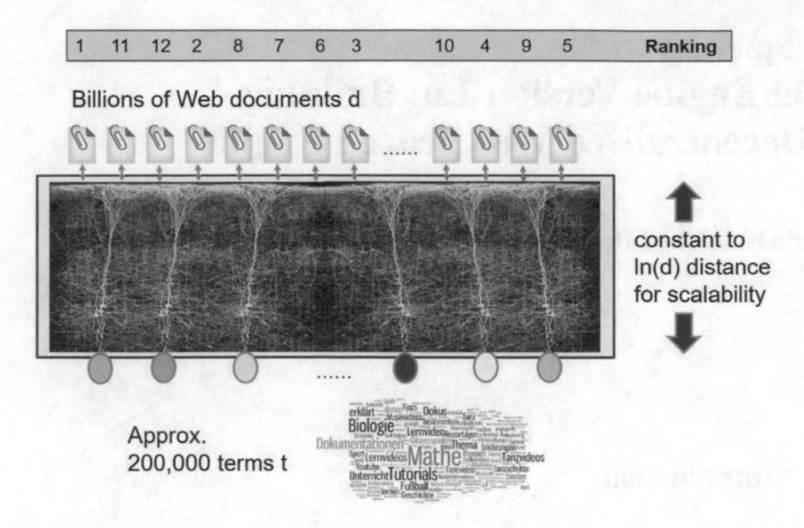

Fig. 3.1 The dimensionality problem of the WWW

However, due to the sheer amount of data to handle and in contrast to the bottom-up approach of using a library catalogue or asking a librarian or human expert in their role as active intermediaries between resources and users for guidance in order to find (more) actually relevant documents, the search engines' approach is less likely to return useful (links to) documents at an instant when the search subject's terminology is not fully known in advance. Furthermore, the web search results are not topically grouped, a service that is usually inherently provided by a library catalogue. Therefore, conducting research using web search engines means having to manually inspect and evaluate the returned results, even though the presented content snippets provide a first indication of their relevance.

From a technical point of view, the search engine's architecture carrying out the mentioned procedures has some disadvantages, too: In order to generate a refindable connection between contents and their locations and to be able to present recent results, the crawlers must frequently download any reachable web pages and thus create and store (multiple) copies of the entire web in their indexes. To achieve a high coverage and actuality (web results should cover contents that have been updated in the last 24 hr with a probability of at least 80%), they cause avoidable network load. Problems get bigger, once the hidden web (deep web) is considered besides regularly accessible HTML (Hypertext Markup Language) pages (surface web), too. Modern web topology models (like the evolving web graph model [2]) emanate from the fact that there are linear as well as exponential growth components, if the overall number of websites is considered. The constant crawling of these components causes especially high network load and their archiving needs a huge amount of storage capacity, too.

This brute-force method of making the web searchable is therefore characterised by a significant overhead for maintaining and updating the indexes. Furthermore, the used technical components like servers and databases are potential targets for cyber-attacks and pose a threat to the system's safety and security as well as for data protection.

As it is necessary to properly address all these problems of centralised web search engines, this paper introduces a new concept along with its technical solutions and infrastructures for future, decentralised web search relying on peer-to-peer (P2P) technology. In order to show that P2P-technology is actually useful in information retrieval tasks, the following section discusses several approaches in this regard before deriving the respective requirements for this concept.

3.2 P2P Information Retrieval

When it comes to using P2P-systems for the purpose of information retrieval, one has to keep in mind that—in contrast to the use case of content delivery—replica of (relevant) documents often do not exist. Thus, it is needed to find the few peers that actually can provide them. Therefore, efficient routing mechanisms must be applied to forward a query to exactly those matching peers and to keep network traffic at a low level. Consequently, a suitable network structure must be set up and adapted in a self-organising manner as well. At the same time, such a network must be easily maintainable.

Some of the most important results in the field of P2P information retrieval (P2PIR) have been obtained in the SemPIR projects [3]. Their goal was to make search for information easier in unstructured P2P-networks. In order to reach this goal, a self-organising semantic overlay network using content-depending structure building processes and intelligent query routing mechanisms has been built. The basic idea of the approach applied therein is that the distribution of knowledge in society has a major influence on the success of information search. A person looking for information will first selectively ask another person that might be able to fulfil her or his information need.

In 1967, Milgram [4] has shown that the paths of acquaintances connecting any two persons in a social network have an average length of six. These so-called small-world networks are characterised by a high clustering coefficient and a low average path length. Thus, the mentioned structure building processes conceived modify peer neighbourhood relations such that peers with similar contents will become (with a high probability) direct neighbours. Furthermore, a certain amount of long-distance links (intergroup connections) between peers with unrelated contents is generated. These two approaches are implemented in order to keep the number of hops needed (short paths) to route queries to matching peers and clusters thereof low. This method is further able to reduce the network load.

In order to create those neighbourhood relations, a so-called 'gossiping' method has been invented. To do so, each peer builds up its own compact semantic profile (following the vector space model) containing the k most important terms from its documents which is periodically propagated in the network in form of a special structure-building request, the gossiping-message. Receiving peers compare their own profiles with the propagated one and

1. put the requesting peer's ID and profile in the own neighbourhood list and
2. send the own profiles to the requesting peer if the profiles are similar to each other.

Also, the requesting peer can decide based on the received profiles which peers to add to its neighbourhood list. Incoming user queries (in the form of term vectors as well) are matched with the local profile (matching local documents will be instantly returned, too), the profiles of neighbouring peers and are forwarded to the best matching ones afterwards. This mechanism differs from the mentioned approach in real social networks: in the technical implementation, the partaking peers will actively route queries from remote peers. In real social networks, people will likely just give the requesting person some pointers to where to find other persons that have the required knowledge instead of forwarding the requests themselves.

In doing so, a semi-structured overlay P2P-network is built which comprises of clusters of semantically similar peers. Additionally, each peer maintains a cache of peers (egoistic links) that have returned useful answers before or have been successfully forwarded queries to matching peers. Furthermore, the network's structure is not fixed as it is subject to dynamic changes based on semantic and social aspects.

Further approaches to P2P-based search engines are available, too. For instance, *YaCy* (https://yacy.net/de/index.html) and the discontinued solution *FAROO* are the most famous examples in this regard. However, although they aim at crawling and indexing the web in a distributed manner, their respective client-sided programs are installed and run on the users' computers. They are not integrated in web servers or web services and thus do not make inherent use of the web topology or semantic technologies for structure-building purposes. Especially, they do not take into account semantic relationships between documents.

3.3 Conceptual Approach

This section introduces the new concept for decentralised web search mentioned in the introduction. Beforehand, important requirements for its realisation are derived from the previous considerations.

3.3.1 Requirements

Based on and in continuation of the foregoing considerations and identified shortcomings of current web search engines, the authors argue that a new kind of decentralised search engine for the WWW should replace the outdated, more or less centralised *crawling-copying-indexing-searching* procedure with a scalable, energy-efficient and decentralised *learn-classify-divide-link* & *guide* method, that

1. employs a learning document grouping process based on a subsequent category determination and refinement (including mechanisms to match and join several categorisations/clusters of words (terms) and documents) using a dynamically growing or changing document collection (the local knowledge base),
2. is based on a fully decentralised, document management process that largely avoids the copying of documents and therefore conserves bandwidth,
3. allows for search inquiries that are classified/interpreted and forwarded by the same decision-making process that carries out the grouping of the respective target documents to be found,
4. ensures that the returned results are 100% recent,
5. returns personalised results based on a user's locally kept search history yet does not implicitly or explicitly propagate intimate or personal user details to any centralised authority and therefore respects data privacy and contributes to information security and
6. returns results without any commercial or other third-party influences or censorship.

Differing from the approaches cited above, the authors intend to build and maintain a P2P-network whose structures are directly formed by considering content- and context-depending aspects and by exploiting the web's explicit topology (links in web documents). This way, suitable paths between queries and matching documents can be found for any search processes. In the next subsections, the respective concept is presented.

3.3.2 *Preliminary Considerations*

In the doctrine of most teachers and based on the users' experience, the today's WWW is considered a client/server system in the classical sense. Web servers offer contents to view or download using the HTTP protocol while every web browser is the respective client accessing content from any server. Clicking on a hyperlink in a web content means to be forwarded to the content, whose address is given in the URL (Uniform Resource Locator) of the link. This process is usually referred to as surfing the web.

Nevertheless, any web server may be regarded as a peer, which is connected to and therefore known by other peers (of this kind) through the addresses stored in the links of the hosted web pages. In such a manner, the WWW can be regarded as a P2P-system (with quite slow dynamics with respect to the addition or removal of peers). However, this system only allows users to surf from web document to web document by following links. Also, these restricted peers lack client functionalities (e.g. communication protocols) offered by web browsers such that there is usually no bidirectional communication between those peers possible (simple forwarded HTTP requests neglected which are mostly initiated by web browsers in the first place).

Moreover, as an integrated search functionality in the WWW is missing so far (the aforementioned restrictions might have contributed to this situation), centralised web search engines have been devised and developed with all their many shortcomings discussed before. These problems will be inherently addressed by the subsequent concept and its implementation.

3.3.3 Implementation Concept

In order to technically realise the mentioned decentralised web search engine, common web servers shall be significantly extended with the needed components for automatic text processing (clustering and classification of web documents and queries), for the processes of indexing and searching of web documents and for the P2P-network management. A general architecture of this concept can be seen in Fig. 3.2.

The concept scheme shows that a P2P-component is attached to standard web server. Its peer neighbourhood is induced by the incoming and outgoing links of local web documents. By this means, a new, fully integrated and decentralised web search engine is created.

In the following implementation-specific elaborations, the P2P-client software 'WebEngine', which follows this concept, is described.

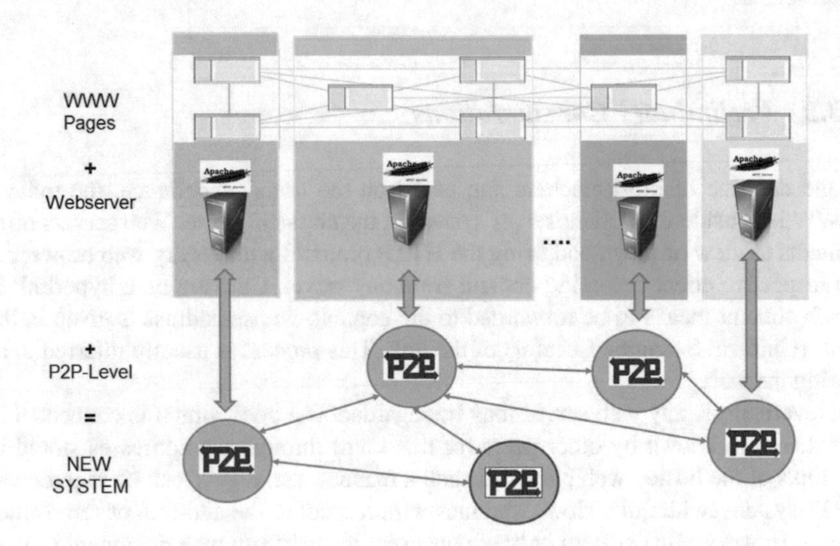

Fig. 3.2 First concept of a decentralised, integrated web search system

3.4 Implementation

As a prototype for this concept, the Java-based P2P-plug-in ‘WebEngine’ for the popular Apache Tomcat (http://tomcat.apache.org/) servlet container and web server with a graphical user interface (GUI) (see Fig. 3.3) for any standard web browser has been developed. Due to its integration with the web server, it uses the same runtime environment and may access the offered web pages and databases of the server with all related meta-information.

The following key points are addressed:

1. A connected, unstructured P2P-system is set up. Initially, the links contained in the locally hosted web pages of the Apache Tomcat server are used for this purpose. Other bootstrap mechanisms as known from [5] and the $PING/PONG$-protocol from *Gnutella* and other P2P-systems may be applied at a later time, too. Note, that
   - HTTP (HTTPS if possible) is used as frame protocol for any communication between the peers.

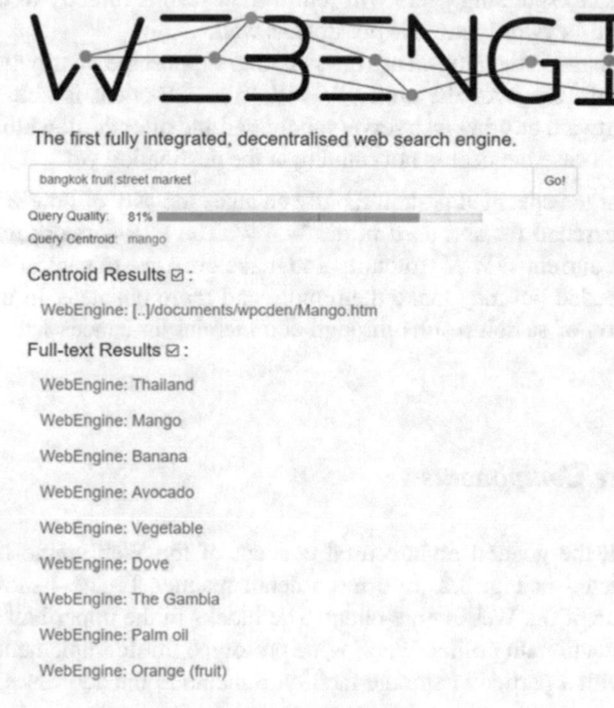

Fig. 3.3 The graphical user interface of the WebEngine

   - A fixed number of connections between the peers will be kept open (although more contactable neighbours are locally stored).
   - Furthermore, a time-to-live (TTL) counter is used to limit the number of forwarded messages.
2. All hosted web documents will be indexed in separate index files after applied stopword removal and stemming.[1] The index is updated after every change in one of the hosted web documents. It acts as a cache to answer incoming queries in a fast manner. However, it would be sufficient—as shown in [6]—to only store and use the centroid terms (single descriptive terms found using preferably large co-occurrence graphs to represent queries and whole texts alike) of local documents and the neighbouring peers' document centroids (their topical environment) in order to be able to route and answer queries properly.
3. The plug-in is able to provide a GUI. In particular, a suitable search page for the requesting user (see Fig. 3.3) is generated.
4. Search results will be generated through a search in the local index files. Queries will also be sent via flooding to all opened connections to neighbouring peers. As they contain a unique message ID, incoming duplicates are discarded. As mentioned before, a TTL counter is applied to limit the number of hops of a query in the network. Responding peers will return their results directly to the originating peer. Multi-keyword search is possible as well.
5. Proliferation mechanisms in the plug-in are integrated to support the distribution of the WebEngine-software over the entire WWW. The P2P-client is able to recognise the peer software on other web servers addressed and offer the download of its own program, in case the peer is not running at the destination yet.

The authors hope that the specified system rapidly changes the way of how documents are accessed, searched for and used in the WWW. The P2P-network may slowly grow besides the current WWW structures and make even use of centralised search engines when needed but may make them more and more obsolete. In this manner, the manipulation of search results through commercial influences will be greatly reduced.

3.4.1 The Software Components

In the previous section, the general architectural concept of the WebEngine has been outlined and depicted in Fig. 3.2. In a more detail manner, Fig. 3.4 shows the software components of the WebEngine-client. The blocks in the upper half of the scheme depict the functionality of the WebEngine prototype (basic implementation) presented so far with a particular storage facility to maintain the addresses of neighbouring peers.

---

[1] In the first version of the P2P-plug-in, indexing is limited to nouns and names as the carriers of meaning.

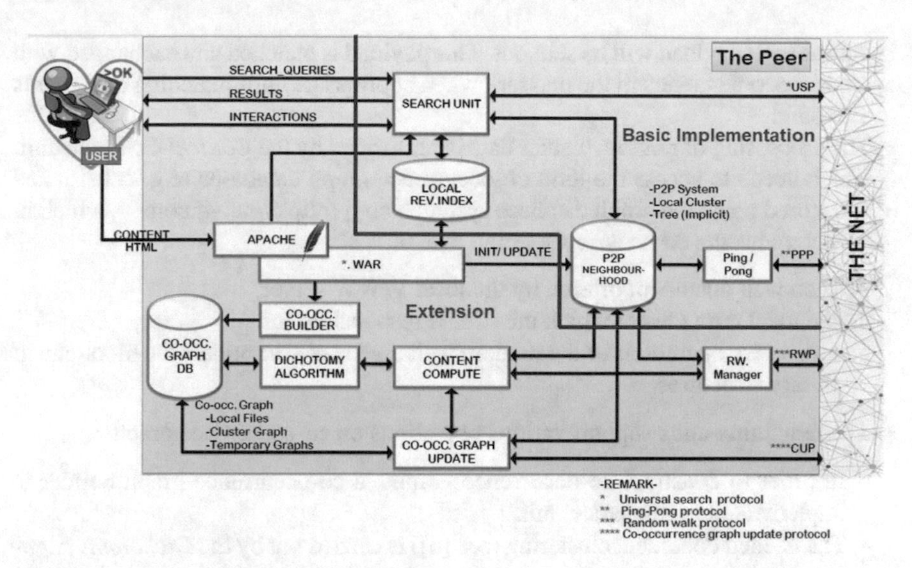

Fig. 3.4 The WebEngine's internal structure

The *Search Unit* is responsible to index local documents as well as to locally answer, forward and handle search requests issued by users. As mentioned above, in the WebEngine prototype, queries will be sent via flooding to all opened connections to neighbouring peers. However, a replacement for this basic procedure in form of a single-message, non-broadcasting, universal search protocol (USP), which forwards the search requests based on the centroid distance measure [7] to the target peer(s), is available (to be activated), too.

In its lower half, Fig. 3.4 presents more advanced features of the WebEngine. As it is the goal to turn the WebEngine into a powerful 'Librarian of the Web' [6], more sophisticated, centroid-based methods for the local management of document collections (their cataloguing, classification and topical clustering), the semantically induced query interpretation and targeted forwarding to neighbouring peers as well as the decentralised construction and maintenance of hierarchical library structures usually comprising a large number of connected peers have been devised and are carried out by these components. The following components have been integrated for this purpose:

- As the decentralised library management is—in contrast to the top-down algorithm presented in [6]—carried out using random walkers, a particularly structured data unit circulating in the P2P-network, in the actual implementation of the WebEngine, a special *RW-Management* unit is added that carries out a special random walker protocol (RWP). Its working principles and methods are described in detail in [8]. Also, the mentioned USP will be additionally extended such that random walkers will not only be used to generate the tree-like library but to perform search operations in it as well. For this purpose, special random walkers with query data as their payload will be sent out. This payload is matched and exchanged with other random walkers in the network when appropriate until matching documents are found.

- The processing of random walker data is performed by the *Content Compute* unit, which needs to access the term co-occurrence graph databases (e.g. constructed and stored using the graph database system Neo4j (https://neo4j.com/)), which in turn contains the co-occurrence graph data of

  1. each web document offered by the local WWW server,
  2. the local term cluster which the peer is responsible for,
  3. temporary operations of the random walkers to build or update the hierarchical library structures.

- The remaining units support various operations on co-occurrence graphs:

  – In order to construct co-occurrence graphs, a co-occurrence graph builder is implemented in a separate unit.
  – The needed document clustering (see [6]) is carried out by the *Dichotomy Algorithm* unit.
  – The exchange of co-occurrence information between peers is supported by the co-occurrence update protocol (CUP) controlled by a respective separate unit.

As the WebEngine makes heavy use of graph databases for the storage and retrieval of co-occurrences, their role is discussed in the next subsection, too.

3.4.2 Graph Databases

When taking a look at the WebEngine's architecture and functionalities from a technological point of view, it becomes obvious that it is necessary to be able to manage large graph structures efficiently and effectively. Graph database systems such as Neo4j (https://neo4j.com/) are specifically designed for this purpose. Also, they are well-suited to support graph-based text mining algorithms. This kind of databases is not only useful to solely store and query the herein discussed co-occurrence graphs with the help of its property graph model, nodes (terms) in co-occurrence graphs can be enriched with additional attributes such as the names of the documents they occur in as well as the number of their occurrences, too. Likewise, the co-occurrence significances can be persistently saved as edge attributes. Graph databases are thus an immensely useful tool to realise the herein presented technical solutions. Therefore, the WebEngine makes especially use of embedded Neo4j graph databases for the storage, traversal and clustering of co-occurrence graphs and web documents.

3.5 Extending the WebEngine

After the launch of WebEngine Version 1.0 in February 2019, the works on it have steadily continued. Now, it is possible to index dynamic web contents and process further document file formats like DOC(X), PPT(X) and ODT besides the originally supported formats PDF, HTML and plain text. A number of performance improvements have been achieved as well. Also, in order to account for confidentiality and integrity, encrypted connections using HTTPS (HTTP over TLS) between the peers are now used by default. Furthermore, numerous GUI adaptations have been conceived and implemented. As an example, it is now possible to switch to another WebEngine instance from entries on the search results page and for administrators to configure the own WebEngine instance from the GUI, too. There is now even a subpage that presents the current instance's statistics such as the number of indexed documents, neighbouring peers, active and completed search processes and the utilisation with e.g. memory-related information. Additionally, a responsive dynamic web page for mobile devices has been designed and provided (see Fig. 3.5).

In order to make centroid terms a more useful tool in text-based search tasks, a concept for a novel associative memory with a ring-like structure [10] storing and managing these terms and their associated contents was proposed. This way, cen-

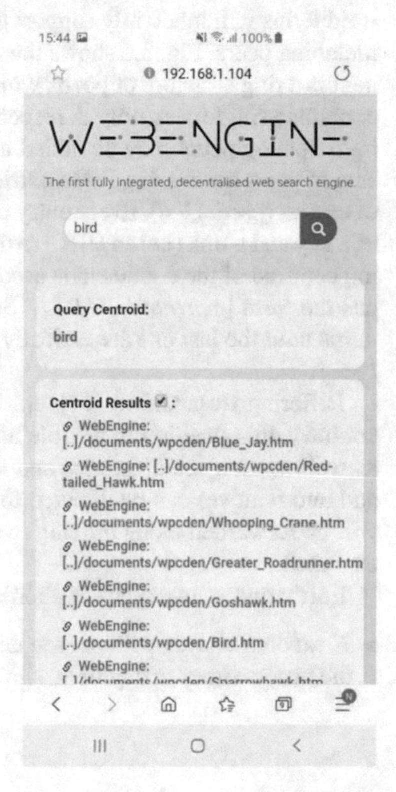

Fig. 3.5 The WebEngine's mobile GUI

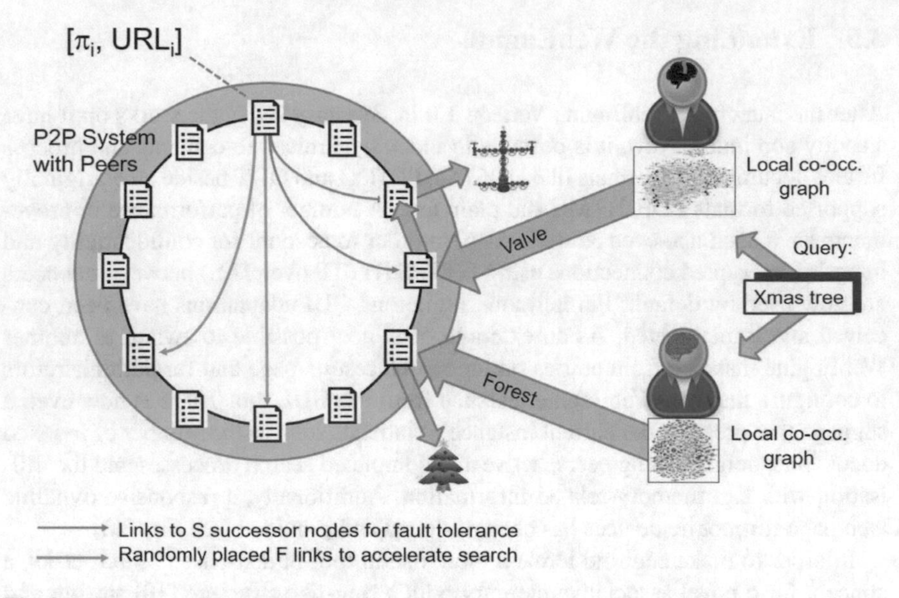

Fig. 3.6 The associative memory's concept

troid terms will inherently support the efficient routing and forwarding of queries to matching peers. Fig. 3.6 shows the design of the proposed system. Its main component is a ring structure of peers, which are running on different machines. It initially contains a single peer, only. A respective management functionality ensures that new, participating peers may be added and—if not needed anymore—be removed. The ring of peers hosts a ring list of entries, whereby every peer can store a larger number of entries (partial list). Every entry represents one text document by a pair consisting of a key and a link (i.e. an URL) to the document. As searchable key, the corresponding centroid of the document is used. In such a manner, the $i-$th entry in the ring list has the form $[centroid_i, URL_i]$. Seen from the first entry $[centroid_1, URL_1]$, all items until the last one are lexically ordered, i.e. $centroid_i \leq centroid_j, \forall i, j$ with $i < j$.

Differing from Chord [11], entries are not assigned to a fixed position. On the contrary, this position is flexible and depends on the number of entries and peers as well as their load. Queries and requests for update operations of items (search, add and remove) can be directed to any peer participating in the ring structure and will be forwarded along the ring to the respective place, where the operation can be executed.

Furthermore, two types of chords are added to each peer of the ring:

- F random chords with chosen destinations from all peers of the ring, shortening the access times to any item, similar to the fingers in Chord.

- S chords connecting any peer with its K nearest neighbours for fault tolerance reasons. Therefore, each peer must mirror the associative memory's contents of those K peers as a backup copy in case one of these peers suddenly leaves the ring without carrying out a proper exit procedure.

Last but not least, every user may access the ring structure by a respective peer service. This service includes

- an individual and user-dependent co-occurrence graph, which allows for the calculation of a centroid to handle a query or document, is built depending on the experience and history of the user and possibly his or her files on the local system.
- a (bounded) broadcast mechanism (TTL 3.4) to locate at least one peer of the ring.
- functions with the respective communication protocol to execute the respective operations on the ring.

Besides this important addition, the search process will now return topically close terms that are semantically related to the entered search query. Actually, those terms are the second-, third-, fourth-best centroid candidates [9] and so on. With the help of those terms, it more easily possible to extend and reformulate the current query. In conjunction with this addition, not only documents with exactly matching centroid terms will be returned, also documents whose centroid terms are close to the query's centroid term will be returned and ranked according to their distance to it. This perimeter search solution will therefore return more matching centroid-based search results.

The newly supported local-only mode helps adminstrators and researchers to let WebEngine run on single computers with no connections to other peers and e.g. analyse topically close document sets in order to support specific investigations at hand. This approach in conjuction with the previously introduced new features has been successfully tested (see Fig. 3.7) on a continuously growing medical corpus consisting of initially 227 English Wikipedia articles covering the most common diseases in the world.

The main finding was that given certain symptoms, the WebEngine instance running in local-only mode was mostly able to identify candidates for diseases causing them. This approach can for instance be used to support physicians to establish differential diagnoses in teleconsultations. Furthermore, it has been implemented in a dedicated medical recommender system incorporating WebEngine's technology.

3.6 Conclusion

This article presented the concept of a novel, decentralised web search engine as well as its P2P-based implementation, called the 'WebEngine'. Its features and software components have been elaborated on in detail. Specifically, it utilises existing web technologies such as web servers and links in web documents to create a decentralised and fully integrated web search system. In doing so, the structure of the generated

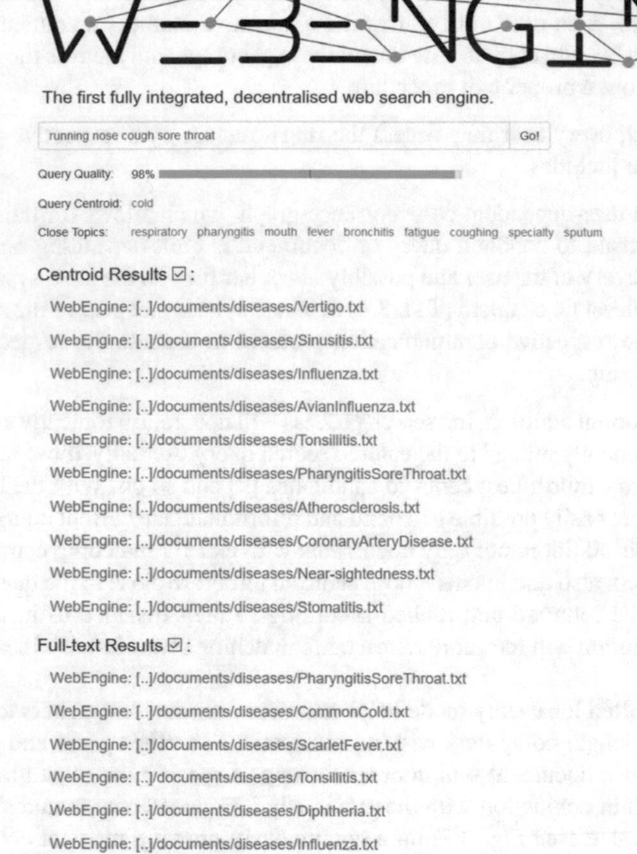

Fig. 3.7 WebEngine's local-only mode

P2P-network is directly induced by exploiting the web's explicit topology. As an extension to the well-known Apache Tomcat servlet container, the WebEngine is easy to install and maintain for administrators. Internally, it makes use of graph-based text analysis techniques. Therefore, the graph database system Neo4j has been used for the persistent storage and retrieval of terms, links between documents as well as the determination of their semantic relations. As such, a decentralised web search system is created that—for the first time—combines state-of-the-art text analysis techniques with novel, effective and efficient search functions as well as methods for the semantically oriented P2P-network construction and management. The basic implementation of the WebEngine has been greatly enhanced by numerous additions which turn it into a modern 'Librarian of the Web'. The WebEngine was made publicly available in early 2019.

References

1. Eberhardt, R., Kubek, M.M., Unger H.: Why Google isn't the future. Really not. In: Autonomous Systems 2015, pp. 268–281. VDI Verlag (2015)
2. Broder, A. et al.: graph structure in the web: experiments and models. In: Computer Networks: The International Journal of Computer and Telecommunications Networking, pp. 309–320. Amsterdam, The Netherlands (2000)
3. Website of the DFG-project 'Search for text documents in large distributed systems' (2009). http://gepris.dfg.de/gepris/projekt/5419460
4. Milgram, S.: The small world problem. Psychol. Today **2**, 60–67 (1967)
5. Kropf, P., Plaice, J., Unger, H.: Towards a web operating system. In: Proceedings of the World Conference of the WWW, Internet and Intranet (WebNet'97), pp. 994–995. Toronto (CA) (1997)
6. Kubek, M.M., Unger, H.: Towards a librarian of the web. In: Proceeings of the 2nd International Conference on Communication and Information Processing, ICCIP '16, pp. 70–78. ACM, New York, NY, USA (2016)
7. Kubek, M.M., Unger, H.: Centroid terms as text representatives. In: Proceedings of the 2016 ACM Symposium on Document Engineering, DocEng '16, pp. 99–102. ACM, New York, NY, USA (2016)
8. Kubek, M.M., Unger, H.: A concept supporting resilient, fault-tolerant and decentralised search. In: Auton. Syst. 2017, Fortschritt-Berichte VDI, **10**(857), 20–31. VDI-Verlag Düsseldorf (2017)
9. Kubek, M.M., Böhme, T., Unger, H.: Empiric experiments with text-representing centroids. Lect. Notes Inf. Theory **5**(1), 23–28 (2017)
10. Unger, H., Kubek, M.M.: An associative ring memory to support decentralised search. Auton. Syst. 2018, Fortschritt-Berichte VDI, **10**(862), 31–45. VDI-Verlag Düsseldorf (2018)
11. Stoica, I., Morris, R., Karger, D., Kaashoek, M.F., Balakrishnan, H.: Chord: a scalable peer-to-peer lookup service for internet applications. ACM SIGCOMM Comput. Commun. Rev. **31**(4), 149 (2001)

Chapter 4
The Brain: WebEngine Version 2.0

Supaporn Simcharoen and Herwig Unger

4.1 Introduction

At present, several resources, including websites, academic papers, articles, news, photographs, and other types of files that are available on the internet, can be easily accessed by search engines. These search engines generally have the main processes: (distributed) crawling, sorted (indexed) copy of the World Wide Web (WWW), and serving results [1, 2]. However, the search results generated from these search engines still lack the ability to consider the respective user's knowledge, despite several techniques had been invented to deliver better search results.

Google, Bing, and Yahoo are the most famous centralised search engines [3]. Because they have a global view of the data collected on the central servers, communications can also be easier. Moreover, they can help to reduce overlapping or duplicate activities. In contrast, the central storages and bandwidth capacity are expensive and may become overloaded when a large content requires greater storage capacity. Furthermore, a single point of failure may happen.

A decentralised search engine is a search system without a central server. The popular decentralised search engines are YaCy [4] and FAROO [5]. Their operations consist of crawling, indexing, and processing user queries which are distributed among several peers. Decentralised search engines are anticipated to solve the problems and the privacy issue found in the centralised search systems. Besides, the system can increase reliability and scalability.

However, the numbers of search results are overwhelming, and most of them may be irrelevant. Moreover, they are without semantic relationships between documents. WebEngine [6], a fully integrated, decentralised web search engine, has taken this into account. The key idea of the WebEngine is that the web servers on the internet are extended by a P2P-search client, which can directly access the proposed contents. The WebEngine system can be easily extended. The connections dynamically add between peers are based on the structure of the existing WWW and the text analysis of specific contents.

H. Unger and M. Kubek, *The Autonomous Web*, Studies in Big Data 101,
https://doi.org/10.1007/978-3-030-90936-9_4

A co-occurrence graph [7] is a powerful underlying information structure of a given text corpus. Co-occurrences are defined as word pairs that appear as direct neighbours or together in a sentence of each document. They are then constructed to produce the co-occurrence graph. When numerous word pairs are inserted, co-occurrence graphs are large graph structures that cannot be easily managed on a single computer. Likewise, the creation of co-occurrence graphs from a corpus needed the text processes, including word segmentation, stemming, and stop-word removal. A significant processing time is required for these complex tasks. Therefore, powerful parallel computing may be necessary.

Peer-to-Peer (P2P) systems have been employed in several applications [8] due to their ability to generate fast, flexible, and fault-tolerant solutions. The P2P networking has been used on scalability issues by distributing resources over several network processes. Every peer appears as both a client and a server. They provide a part of the system resources. All peers are free to join or leave the system [9].

Therefore, in a large network, the search for specific information may become a tedious task which is usually resolved by [8]:

- Requested messages are broadcasted to the entire system, which leads to an overloading of the communication channels. The number of messages exponentially increases with each step. (e.g., in GNUTELLA [10]) or
- Requires a long time for serial search that can be improved by establishing numerous copies of the accessible contents (e.g., in FREENET [11]) or
- Building tree or ring-like structures that can be searched efficiently. However, the system may be loaded with respective management overhead (e.g., in CHORD, CAN, TAPESTRY [12]).

When building large co-occurrence graphs, several computers are required. Moreover, efficient load balancing is also needed, which cannot be delivered by the above-cited approaches. Hence, a fully decentralised co-occurrence graph management system under the project name the web search engine 'TheBrain', is a potential solution to these problems.

TheBrain, a second version of the WebEngine, works on several peers without a controlling authority and can be distributed across the networks. Specific knowledge of the users is applied to operate and improve the searching tasks. Two types of graph databases are created: local co-occurrence graphs (the existing user knowledge) and one global co-occurrence graph. Significant contributions of TheBrain are:

- The system can be used without a central server, which manages the global co-occurrence graph.
- The co-occurrence graph is constructed in a parallelised manner. Every peer can contribute the resources.
- The system offers a fully decentralised and brain-like method to recreate lost semantic connections needed for routing messages like queries.
- The distributed computation will manage not only the calculation of graph operations but also the distribution of memory capacity(load) and network traffic.
- Load distribution strategies are added to attain a good performance.

Furthermore, an oblivion mechanism of learning and forgetting is applied. Facts and relations presented in a repeating manner are learned, whereas facts and relations that are not presented or used for a long time will be forgotten. The verification of text-based information and the detection of irregular and regular changes of the collective knowledge over time (e.g., attained by fake news) will be studied in a different paper.

4.2 Building TheBrain

4.2.1 Concept

In a distributed management of the co-occurrence graph $G = (W, E)$, each peer manages a sub-graph G_i of the overall co-occurrence graph G by storing them with a subset of its words (nodes) and the respective edges. As a result of this, two classes of the edges are found: the edges of the global co-occurrence graph hosted on the same peer and one that hosted on different peers. Therefore, all words on the global co-occurrence graph G have to carry unique identifiers and some information (e.g., IP addresses) of the peer, where they are hosted or where the endpoints of an edge are situated.

Figure 4.1 displays one of the problems related to distributed management of the global co-occurrence graph G. It is not easy to decide whether an added word is already a member of the co-occurrence graph G. Hence, searching for specific information becomes a tedious task, as described above. Moreover, the master computer, as shown in Fig. 4.1, indicates that distributed hash tables are not a satisfying solution since it generates unnecessary bottlenecks that require additional management overhead as well as having the potential to be a source of failure.

TheBrain, a purely distributed P2P system, which facilitates a distributed management of the co-occurrence graph $G = (W, E)$, is proposed (see Fig. 4.2). TheBrain will eliminate the master computer from the system and reduces the bottlenecks occurrences.

TheBrain, a new platform that learns how to process information, can work efficiently by replicating how the human brain works. It differs from the WebEngine [6], including the search process of WebEngine due to an extended Chord-like ring structure which may lead to the load balancing problem. WebEngine can be overloaded by numerous users trying to access the same data item at the same time. On the other hand, the search process of TheBrain does not use structural overlays but relies on routing queries with semantic considerations. Moreover, the management of the ring structure is carried out by automatically selected peers in WebEngine. In contrast, all peers in TheBrain carry management tasks that mainly affect how load balancing is carried out. Nevertheless, an essential difference is that in TheBrain, the peers crawled the entire indexed web. In WebEngine, only the peers' local contents have been indexed. In addition, concerning fault tolerance, TheBrain offers a new fully

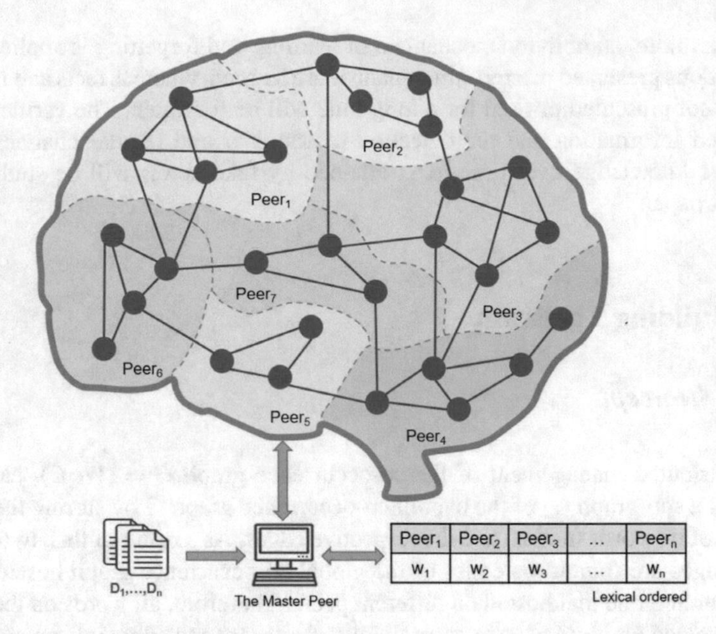

Fig. 4.1 Building a global, distributed co-occurrence graph

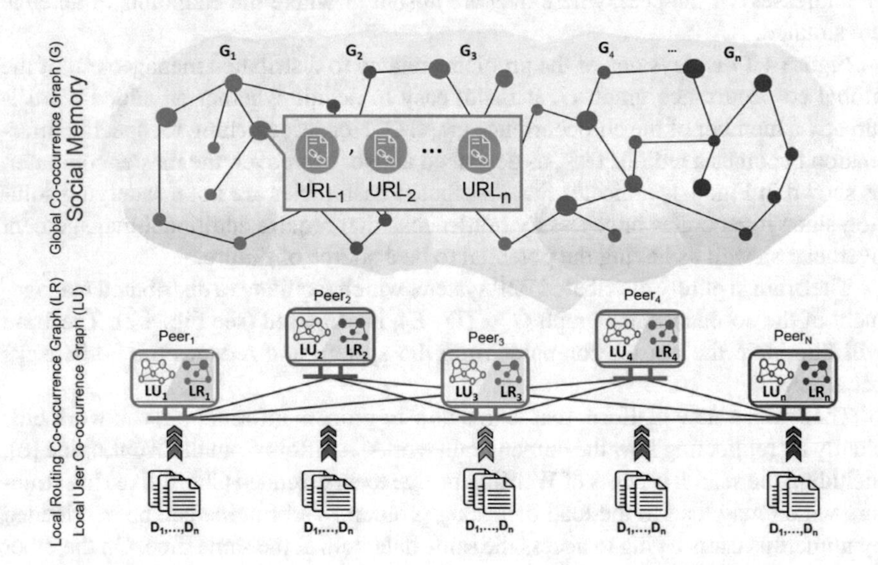

Fig. 4.2 Overview of TheBrain

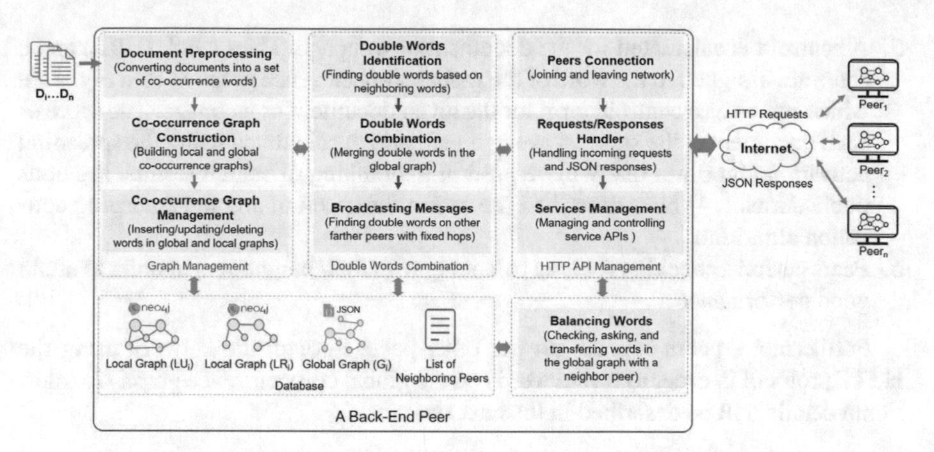

Fig. 4.3 The system flow processes of TheBrain

decentralised and brain-like method to recreate lost semantic connections needed for routing messages like the queries.

In TheBrain, the distribution of the global co-occurrence graph G begins from a peer. Additional peers are free to connect or leave the system. A user i can insert the local documents or specify interest contents in the WWW. Then, the integrated web spider can work on the user's behalf to extend the respective co-occurrence graphs. Besides, a web crawler then downloads and indexes contents of the internet by automatically following a link from a known page to a new page. Figure 4.3 shows the system flow processes of TheBrain. The contents of local documents or webpages are pre-processing and co-occurrence extraction, which occurs in all peers. Consequently, their co-occurrences are concluded, connected, and collected into the global co-occurrence graph G_i, the local user co-occurrence graph LU_i, and the local routing co-occurrence graph LR_i. The processing of documents in order to build the co-occurrence graphs consists of the following steps:

1. The content of each local document or webpage is read sentence by sentence. Each sentence is formed as a composition of $n = |S_i|$ well-ordered co-occurrence $(w_a, w_b)_i$, after stemming and stop-words removal is applied.

$$S_i = \{(w_a, w_b)_i, \ldots, (w_a, w_b)_n\} \tag{4.1}$$

2. Co-occurrences (w_a, w_b) of each sentence S_i are processed for insertion or updating into the local co-occurrence graphs LU_i.
3. Co-occurrences (w_a, w_b) of each sentence S_i are processed by using a minimal spanning tree (MST) of Prim algorithm [13] to reduce redundant edges, getting a suitable graph, and then they are inserted or updated into the local routing LR_i and global co-occurrence graphs G.
4. Words on the global co-occurrence G are checked for double-words.

5. A centroid is calculated for the document or webpage. Then a link (URL) and a score are assigned to the word of the global co-occurrence graph G (see Fig. 4.2), which serves as a centroid term for the given document or webpage. The score of each document is the shortest average path length calculated using the spreading activation algorithm that will be used in the ranking of search results. Previous publications [14] have described the centroid extraction and the spreading activation algorithm.
6. Peers perform checking for load balancing with their neighbours in order to attain good performance.

Furthermore, peers connect several other peers through the network using the HTTP protocol in order to locate words on the global co-occurrence graph G. Additional details will be described in the next sections.

4.2.2 *The Mechanism of Locating Words on TheBrain*

The contents of local documents or webpages are assigned by the user i. These contents are processed, the local co-occurrences graphs LU_i and LR_i are built. Their co-occurrences are then included in the global (distributed) co-occurrence graph G. Before each co-occurrence (w_a, w_b) is inserted into the global co-occurrence graph G, it is verified as follows:

Check 1: One of the two words w_a and w_b or both already exist on the global co-occurrence graph G_i:

1. If both words w_a and w_b are found, then the respective edge is added, or the weight of an existing edge is updated to the global co-occurrence graph G.
2. If words w_a or w_b are not found, then Check 2 is performed.

Check 2: One of the two words w_a and w_b or both are endpoints of the edges on the global co-occurrence graph G_i:

1. If both words w_a and w_b are found, then the respective edge is added, or the weight of an existing edge is updated to the global co-occurrence graph G.
2. If the words w_a or w_b are not found, then Check 3 is performed.

Check 3: One of the two words w_a and w_b or both already exist on the local co-occurrence graph LU_i:

1. If both words w_a and w_b are found, the path p in LU_i (see Fig. 4.4) can also be found on G by starting from the word which is an endpoint of G_i. Follow the path p from this endpoint on G_i will give a route towards the second word w_a or w_b on the global co-occurrence graph G. Then use this path p to send the message to the remote peer to add the respective edge or update the weight of an existing edge on the global co-occurrence graph G.
2. If the words w_a or w_b are not found, then Check 4 is performed.

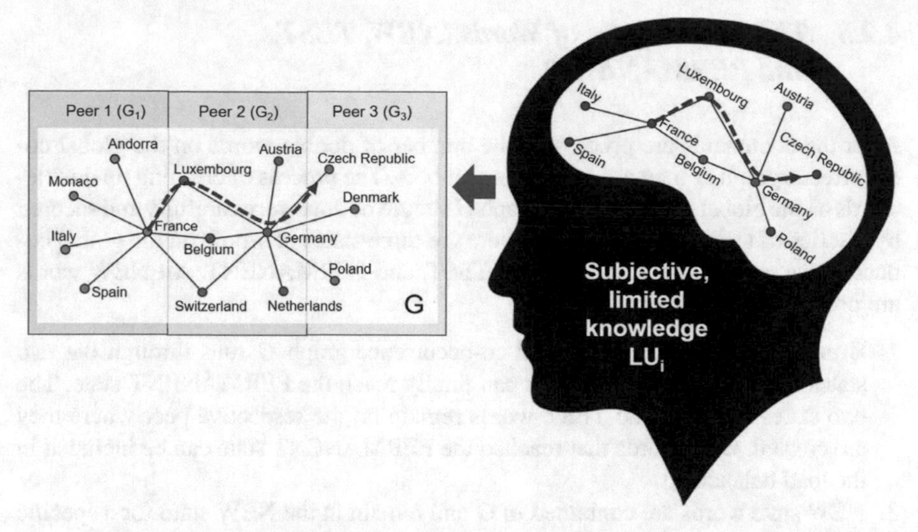

Fig. 4.4 The idea of the local routing method

Check 4: One of the two words w_a and w_b or both already exist on the global co-occurrence graph G_i of neighbouring peers. Subsequently, the requests will be sent to neighbouring peers to locate the words w_a or w_b. If one of the two words w_a and w_b or both are found, then the respective edge is added, or the weight of an existing edge is updated to the global co-occurrence graph G.

If the words w_a or w_b are not found in all cases, then any of the NEW words will be inserted into the peer where it appears for the first time. Next, corrections are executed, during the routing, if a different routing indicates a double existence of one word to the same target. In this case, non-homonym words can be combined. Homonyms such as "mouse" may refer to a mouse, a type of rodent or a mouse, a small handheld electronic device.

In Fig. 4.4, a user formulates a query word q to which a path p from any known word w_j with $p = \{w_j, w_2, \ldots q\}$ is known. This path p also has to exist in the global co-occurrence graph G. Consequently, the user can find the word on the local machine or any neighbour where a word w_j is stored. At the same time, whenever a path $p = \{w_j, w_2, \ldots q\}$ with $w_j \in W_i$ can be found in LU_i, the path p also represents an entire routing path on the global co-occurrence graph G, from w_j towards the storage place of q.

4.2.3 The Three States of Words (NEW, TEST, and PERMANENT)

After the documents are processed, the number of double-words on the global co-occurrence graph G may occur in large numbers. The process of cleaning up double-words on the global co-occurrence graph G should be done very carefully and secured by a series of tests in a given time. There are three states of words on the global co-occurrence graph G, including NEW, TEST, and PERMANENT. All NEW words are processed as follows:

1. Every NEW word on the global co-occurrence graph G runs through the two states NEW and TEST before it can finally reach the PERMANENT state. The two states of NEW and TEST words remain on the respective peer where they are created. Only words that reached the PERMANENT state can be included in the load balancing.
2. NEW-state words are contained in G and remain in the NEW state for a specific time.
3. The percentage p is used as a threshold to ensure that two words are not homonyms before they are combined. Every NEW word w_{new} is checked for double-words in four steps as follows (see Fig. 4.5):

   a. Starting from the word w_{new}, all neighbouring words of w_{new} in the global co-occurrence graph G are requested.
   b. All neighbouring words are checked to determine whether or not their neighbouring words are the same as w_{new}.
   c. All neighbouring words send their respective responses back to the word w_{new}.
   d. The percentage of the same words $p_{w_{new}}$ is calculated by

   $$p_{w_{new}} = \frac{n}{N} \times 100, \tag{4.2}$$

   where n is the number of shared neighbours of each same word that is found, and N is the number of all neighbours of the word w_{new}. If $p_{w_{new}} \geq p$, the two words are assumed to be the same words, i.e., homonyms, then they are combined. Next, the state of the original word w_{new} is changed to PERMANENT.

4. After a given time, the NEW state has elapsed, the word changes its state from NEW to TEST. The duration of the TEST state is significantly longer than the NEW state. The length of the time depends on the size of the network. For checking double-words, all TEST words are processed as follows:

   a. In each (time) step, a word is randomly chosen from all TEST words and defined as a destination word. Then, a message is generated for a search request which contains the following information:

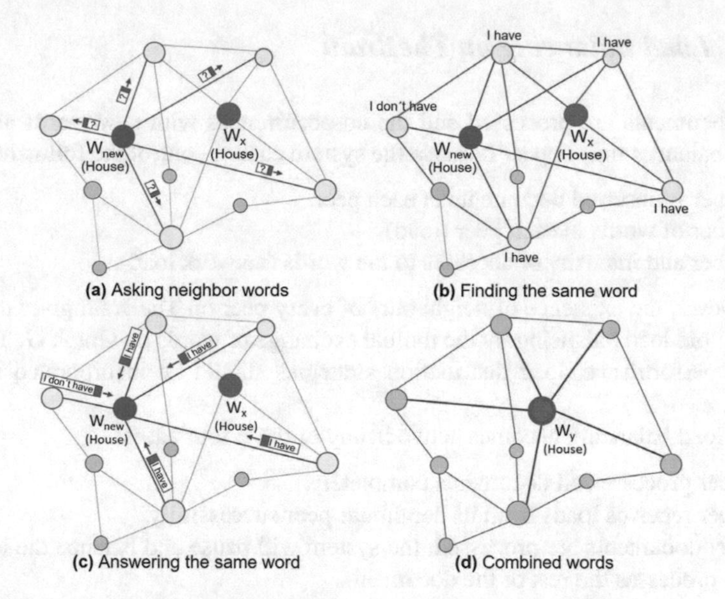

Fig. 4.5 The steps of checking double-words

- IP address of the source peer,
- Destination word: a randomly picked TEST word for checking double-words,
- List of checked peers: the list of IP addresses of peers that are checked for the destination word,
- Number of hops: the number of peers that the message has traversed,
- Maximum number of hops.

b. If the number of hops is equal to the maximum number of hops, then this message is dropped.
c. The message will be forwarded to a neighbour peer (a peer who never get this message before), which is randomly chosen from all neighbour peers.
d. If the destination word is found in a peer, the two words are combined. The status of the word is changed to the PERMANENT state.
e. If the destination word is not found in a peer, the number of hops increases, the IP Address of this peer is added to the list of checked peers, and go to b).

After executing all processes, it can be assumed that the global co-occurrence graph G does not have the double-word. Each word is represented exactly once in the global co-occurrence graph G.

4.2.4 Load Balancing on TheBrain

After documents are processed and the co-occurrences with new words are built, load imbalances may appear because the system cannot control the following:

- Number of inserted documents in each peer.
- Number of words in each peer (load).
- Number and intensity of accesses to the words (network load).

However, the existence of neighbours of every peer on TheBrain gives a chance for a simple load balancing by the mutual exchange of words on Graph G. To attain a good performance, load distribution strategies should be incorporated into the system.

The load balancing becomes active in any of these four cases:

1. A peer processes all documents completely.
2. A peer receives loads from its neighbour peer successfully.
3. Many documents are processed; the system will pause and balance the load and then processes the rest of the documents.
4. A new peer joins TheBrain.

The steps of the loads balancing are described as follows:

1. Each peer starts to check the loads (words) with its neighbours indefinitely, using a random sequence. During the checking and transferring, the participating peers will be LOCKED to prevent access by others.
2. If a peer handles β percentage more words than a neighbour peer, this peer will transfer PERMANENT words to the neighbour peer to equalise the loads.
3. After the transfer of words, the information of its adjacent edges will be adapted on both endpoints. The transferred words and edges are updated into the local co-occurrence graphs LU_i of the target peer(s).
4. Finally, the participating peers are changed status to UNLOCK mode; both peers can be accessed by others or repeatedly request balancing load with other peers.

With these requirements, fast distributed calculations on the co-occurrence graphs with a text corpus will be possible. The results of an implementation of the proposed concepts are described in the next sections.

4.3 Experimental Results

4.3.1 Goals

The given concepts will be evaluated using experiments in order to determine the feasibility and efficiency of TheBrain: the second version of the decentralised web search engine, called WebEngine. The system allows the user to insert local documents or specify interest contents in the WWW. The local co-occurrence graphs and the global co-occurrence graph are built in different numbers of peers. The goals of these concepts are:

1. The P2P system (TheBrain) can be used without a centralised server to manage the global co-occurrence graph.
2. The co-occurrence graphs work parallelly; every peer can contribute its resources to the system.
3. A significant speedup can be achieved.
4. The removal of the double-words problem can be resolved by three states of words (NEW, TEST, and PERMANENT).
5. The load balancing algorithm in each peer can react if an imbalance is detected in the P2P network.

To prove this, as above indicated goals, the experiments will be described below.

4.3.2 Experimental Setup

The experiments were performed under the assumption that the goals of the concepts have been achieved. To ensure that the concepts perform well, documents were inserted, co-occurrence graphs were created, double-words were checked, and load balancing between peers was carried out. TheBrain was started with one peer at the beginning. Then, other peers connected by acknowledging only one existing peer (IP address and Port) on TheBrain. The two datasets used for creating the co-occurrence graphs were as follows:

1. 10 documents from the travel topic; this dataset was inserted into the first peer. All words were immediately set to the PERMANENT state necessary for the load balancing when a new peer joined TheBrain.
2. 1000 documents (200 documents per topic) that cover various topics, including arts, cars, computers, leisure, and sports. This dataset was used to test with five experiments that differed in the number of peers on TheBrain (5, 10, 15, 20, 25 peers).

4.3.3 Results and Discussion

The experiments demonstrated that the proposed concepts had been successfully applied to build co-occurrence graphs with different numbers of peers. The results are presented in Figs. 4.6, 4.7, 4.8, 4.9, 4.10 and 4.11.

The distribution of words on the global co-occurrence graph G started from a distributed system with one peer. The number of peers increased, and the load balancing distributed the load to all peers joining TheBrain, as shown in Fig. 4.6. After the documents were inserted to peer $P1$, the co-occurrence graph was built. The

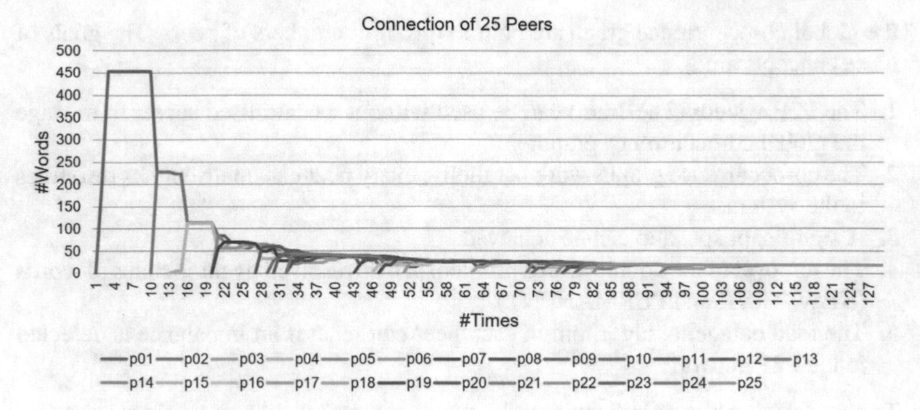

Fig. 4.6 25 Peers connecting to TheBrain

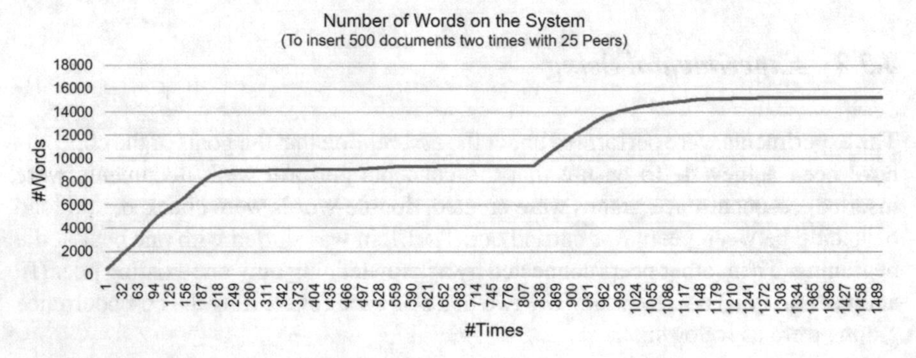

Fig. 4.7 Total number of words on TheBrain

number of new words was 454. All new words were set as PERMANENT words. After that, a second peer $P2$ was connected to TheBrain and became a neighbour of peer $P1$. Next, the load balancing transferred the load (words) to peer $P2$. Over time, peers $P3 - P25$ joined TheBrain, respectively. After peers joined TheBrain, the load balancing distributed the load (words) among peers until a balance was achieved.

After that, the sets of documents were inserted at two times with 25 peers (time = 0 and time = 811). Each set of documents consisted of 500 documents, 20 documents per a peer. An increase in the number of words over time is presented in Fig. 4.7. The number of words on TheBrain significantly grew up at the beginning. After the first set of documents was processed in every peer, words were successfully included in the global and local co-occurrence graphs. The second set of documents was then added and further increased the number of words on TheBrain. However, later increases of words on the global co-occurrence graph happened more steadily over time.

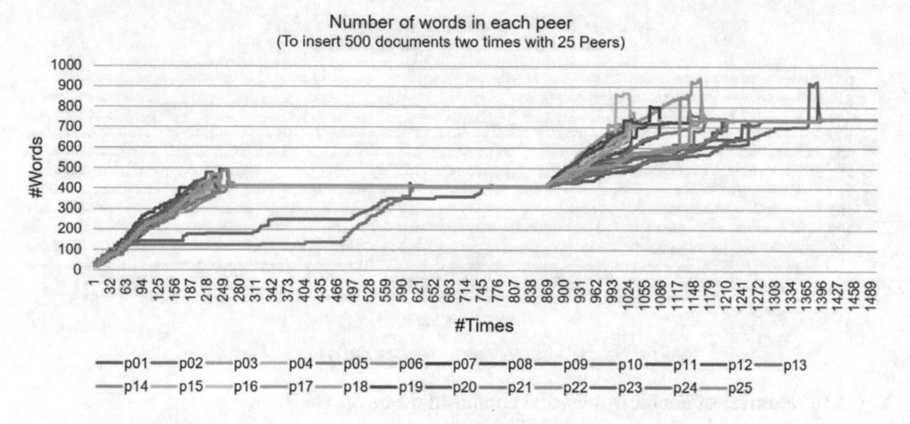

Fig. 4.8 Number of words in each peer on TheBrain

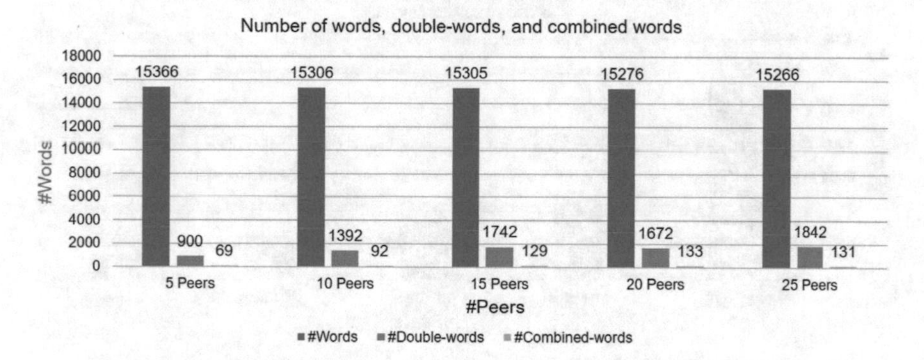

Fig. 4.9 Number of words, double-words, and combined words on TheBrain

During the insertion of documents two times, each document was processed in the peer in which it was added. For every word, the existence of double words was checked in all peers. The words and relations between words were added to the co-occurrence graphs. The number of words in each peer gradually increased over time. After documents were inserted completely in any peer, the peer started to run the load balancing process. As the number of words in each peer increased, the co-occurrence graphs grew gradually, and peers balanced their loads with neighbour peers towards the end of both periods (time = 784 and time = 1432), see Fig. 4.8.

Figure 4.9 presents a comparison of the number of words, the final number of double-words, and the number of combined words with five experiments (5, 10, 15, 20, 25 peers). After 1000 documents were inserted for each experiment, words and relations between words were added into the co-occurrence graphs. The five experiments showed minimal differences in the number of words, the final number of double-words, and combined words. Overall, when the number of peers grew, the number of detected double-words and combined words were also increased.

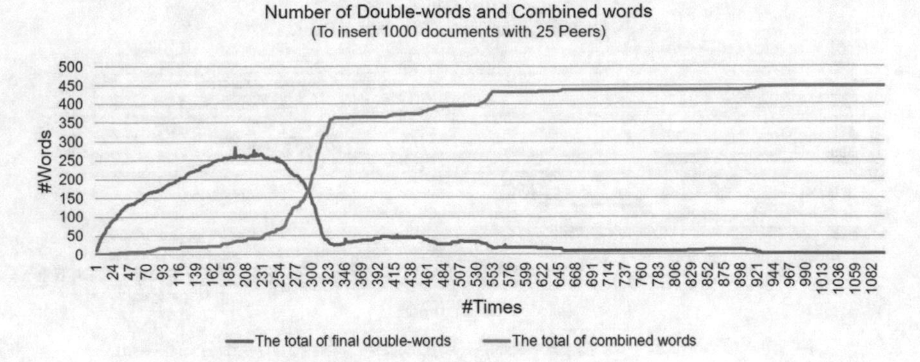

Fig. 4.10 Number of double-words, and combined words on TheBrain

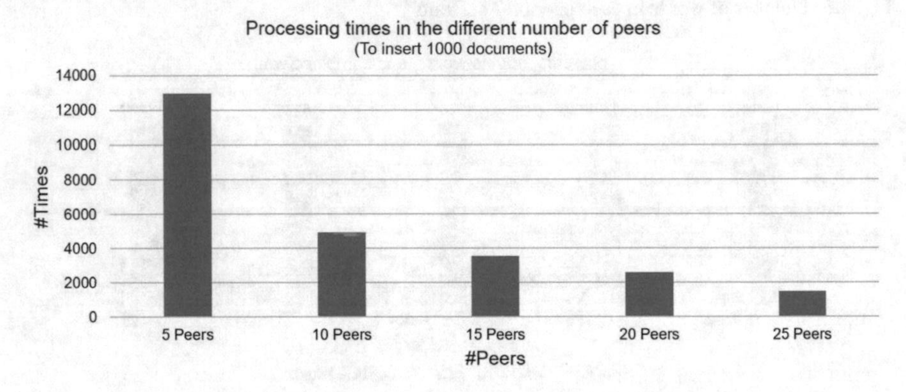

Fig. 4.11 Processing times in the different number of peers on TheBrain

Due to the large number of double words occurring on TheBrain (see Fig. 4.9), Fig. 4.10 shows a decrease in the number of double-words after the documents were inserted completely. In the beginning, the words were added, they remained in the NEW state for a given period. Then, double-words were checked and combined using the percentage threshold p to ensure that two words were not homonyms. After that, the words' state was changed from NEW to TEST, double-words were rechecked and combined using the same-word-request protocol on TheBrain. Finally, the state of the combined words was changed to PERMANENT. Thus, the transition from NEW to TEST and PERMANENT could happen for every word at different points in time.

Furthermore, a decrease in the processing time when inserting 1000 documents into 5, 10, 15, 20, and 25 peers on TheBrain is presented in Fig. 4.11. The rising number of peers on TheBrain led to a decline in the processing times. Therefore, from these experiments, a more significant number of machines processed documents faster than fewer machines.

4.4 The Web Search Engine TheBrain

A fully decentralised co-occurrence graph management system under the project name of a search engine 'TheBrain' based on a peer-to-peer (P2P) technology, has been developed. TheBrain can produce semantic search results related to the search query context. The possible relevance of search results are retrieved based on the co-occurrence graphs.

The processes for searching the documents are shown in Fig. 4.12. When a user requests a search query from the system, the text-representing centroid of this query was calculated in the local co-occurrence graph. Then, a path directly to the word (centroid) on the global co-occurrence graph was returned. Next, the mapped documents (URLs) with the word (centroid) ranked using the score were presented to the user. As previously mentioned, the score is the shortest average path length calcu-

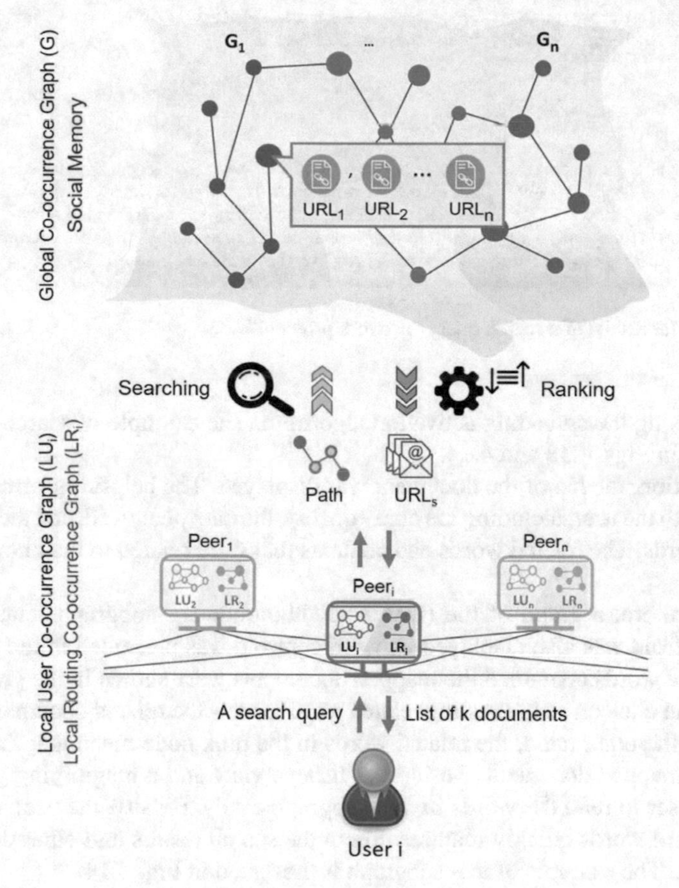

Fig. 4.12 Document searching processes

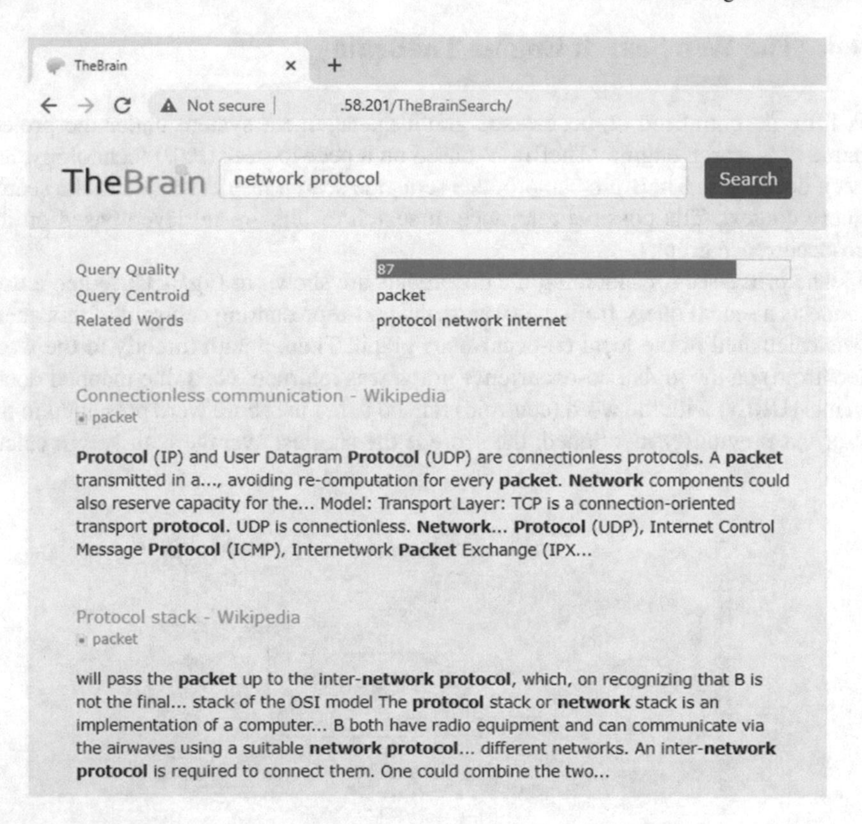

Fig. 4.13 The results of a search query "network protocol"

lated by using the spreading activation algorithm. The example of search results is presented in Figs. 4.13 and 4.14.

In addition, the list of the document was displayed. The helpful information was presented to the user, including the query quality, the query centroid, and the topically related words. The related words had contents that were related to the knowledge of the user.

Furthermore, a graph of the (word) neighbourhoods—subgraph related to the query centroid was also displayed. A word centroid was presented in the red node. The related words contained the mapped documents were shown in the green node; the user can click on an interesting related word to view the related documents of that word. On the other hand, the related words in the pink node mean that they do not have any mapped document. Besides, A fisheye view and a magnifying glass may help the user to read the words in this subgraph easily. Finally, the user would get more related words quickly than reading in the search results and other documents like books. The example of this subgraph is presented in Fig. 4.14.

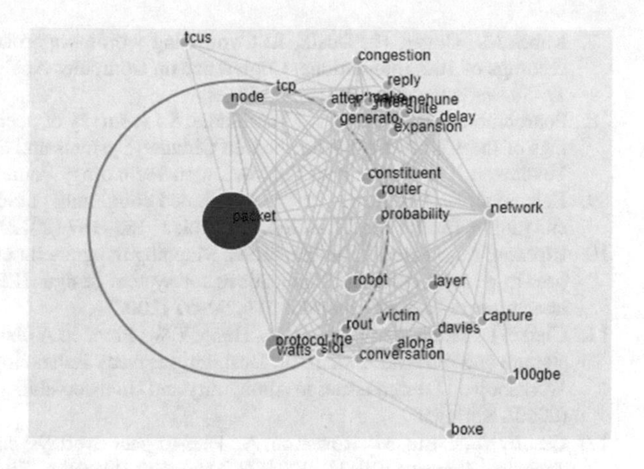

Fig. 4.14 A graph of the (term) neighbourhoods

4.5 Conclusion

This article presents the concepts and implementation details of TheBrain, the second version of the decentralised web search engine, called WebEngine. According to the results, the concepts worked on different numbers of peers. Increasing the number of peers helped to reduce the processing time. When documents were added, words and relations between words were inserted into the co-occurrence graphs. All new words were also checked for double words. Furthermore, a simple load balancing mechanism was activated, allowing equal distribution of loads (words) on the global co-occurrence graph among all participating peers. Overall, the decentralised search engine under the project name the search engine 'TheBrain' could perform semantic search results related to the search query context. Finally, the proposed concepts could be later used effectively by online communities.

References

1. Eberhardt, R., Kubek, M., Unger, H.: Why google isn't the future. really not. In: Autonomous Systems 2015, pp. 268–281. Mallorca, Spain (2015). Fortschritt-Berichte VDI
2. Brin, S., Page, L.: The anatomy of a large-scale hypertextual web search engine. In: Computer Networks and ISDN Systems of 7th International World Wide Web Conference, pp. 107–117. Brisbane, Australia (1998). Elsevier B.V
3. Search engine market share worldwide. https://gs.statcounter.com/search-engine-market-share. Accessed 8 Apr 2021
4. Yacy search engine software. http://www.yacy.de/de/index.html. Accessed 28 Mar 2021
5. Faroo. http://www.faroo.com/. Accessed 28 Mar 2021
6. Kubek, M., Unger, H.: The webengine—a fully integrated, decentralised web search engine. In: Proceedings of the 2nd International Conference on Natural Language Processing and Information Retrieval, pp. 26–31. Bangkok, Thailand (2018). Association for Computing Machinery

7. Kubek,M., Unger, H., Dusik, J.: Correlating words—approaches and applications. In: Proceedings of 16th International Conference on Computer Analysis of Images and Patterns, pp. 27–38. Valletta, Malta (2015). Springer
8. Pourebrahimi, B., Bertels, K., Vassiliadis, S.: A survey of peer-to-peer networks. In: Proceedings of the 16th Annual Workshop on Circuits, Systems and Signal Processing, pp. 570–577, Veldhoven, The Netherlands (2005). Dutch Technology Foundation
9. Kahanwal, B., Pal Singh, T.: The distributed computing paradigms: P2p, grid, cluster, cloud, and jungle. Int. J. Latest Res. Sci. Technol. **1**, 183–187 (2012)
10. Ripeanu, M., Foster, I., Iamnitchi, A.: Mapping the gnutella network: properties of largescale peer-to-peer systems and implications for system design. IEEE Internet Comput. J. (special issue on peer-to-peer networking) **6**, 50–57 (2002)
11. Clarke, I., Sandberg, O., Wiley, B., Hong, T.W.: Freenet: A distributed anonymous information storage and retrieval system. In: Designing Privacy Enhancing Technologies of International Workshop on Design Issues in Anonymity and Unobservability Berkeley, pp. 46–66. CA, USA (2000). Springer
12. Castro, M., Costa, M., Rowstron, A.: Peer-to-peer overlays: structured, unstructured, or both? Technical Report MSR-TR-2004-73, Microsoft Research, Tech., (2004)
13. Prim's minimum spanning tree. https://www.studytonight.com/data-structures/prims-minimum-spanning-tree. Accessed 08 Apr 2021
14. Kubek, M., Böhme, T., Unger, H.: Spreading activation: a fast calculation method for text centroids. In: Proceedings of the 3rd International Conference on Communication and Information Processing, pp. 24–27. Tokyo, Japan (2017). Association for Computing Machinery

Chapter 5
WebMap: A Concept for WebEngine Version 3.0

Georg P. Roßrucker and Herwig Unger

5.1 Vision

With the continued increase of available content on the world wide web (WWW),[1] the need for efficient tools that support web navigation solidifies. As discussed in [1], a small number of predominant, centralized search engines are among the most popular tools to satisfy users' *informational needs*.

To overcome this dependency, in our mind, every instance offering content to the WWW should also contribute to the search and navigation in it. We achieve this by utilizing a peer-to-peer (P2P) network, in which web servers act as peers, sharing information necessary for decentralized routing and search. This P2P approach was originally introduced in the initial WebEngine [2] and utilized in TheBrain [3], too. Figure 5.1 depicts the idea of creating a P2P network between web servers, otherwise decoupled from each other.

The aforementioned approaches rely on P2P background tasks, e.g., to maintain a global graph, and data storage, as well as on-demand tasks like routing, and the processing of search queries. As a consequence, the results generated by one instance (peer/server) may vary in consistency and processing time, depending on the availability of other instances. If any of these is (temporarily) unavailable, the P2P network needs to re-distribute graph data, causing a loss of references across the whole network. Re-discovering peers and re-establishing references result in high expenses for routing and overall load. Besides the mutual reliance of peers, the on-demand processing of search requests is based on graph database operations, which do not only occupy valuable resources but are also expensive in terms of processing time.

With our new approach we intend to mitigate the mutual reliance in on-demand processes. Following the example of sitemaps,[2] we introduce *shadow* and *cluster files*, which establish a static linking structure on the WWW, as sitemaps do in the

[1] http://www.internetlivestats.com/total-number-of-websites/.

[2] https://www.sitemaps.org/index.html.

H. Unger and M. Kubek, *The Autonomous Web*, Studies in Big Data 101,
https://doi.org/10.1007/978-3-030-90936-9_5

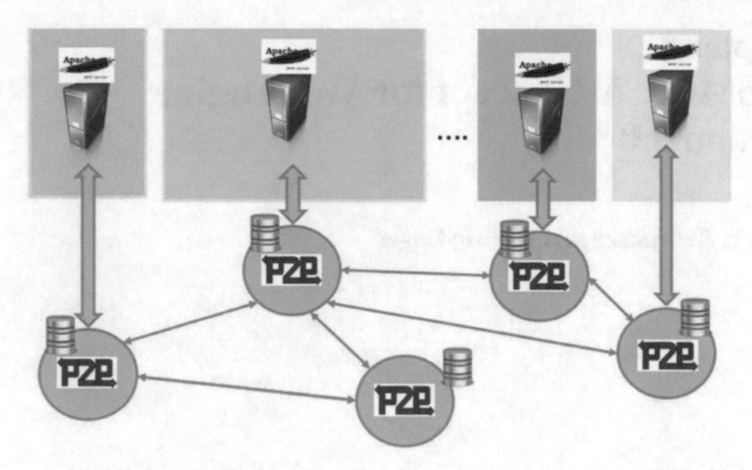

Fig. 5.1 Otherwise decoupled web servers form a P2P network, extending the connectivity of the WWW

context of single domains. While a P2P network still plays a role in managing these static files and performing other background tasks, on-demand tasks will be based on static files, independent from the P2P network. We expect that this will significantly reduce load and increase performance.

Our objective for the proposed WebEngine 3.0 is to establish static shadow and cluster files, and develop algorithms to support the following applications:

1. Forming clusters, by assigning websites to the same category, expressed by a text representing term, e.g., TRC. A cluster contains hyperlinks to websites with similar topics and is made available to the WWW.
2. Establishing an augmented web graph by creating shadow files for each web page hosted by a server. A shadow file contains links to other relevant resources, i.e., web pages, and cluster files.
3. Proposing relevant resources covering related information to users, by generating link recommendations based on user-stated search queries, or other inputs like web pages or documents.

In analogy to sitemaps, we name the network of connected shadow and cluster files *WebMap*. The main challenges will be to manage and maintain the static files over time and develop an efficient routing algorithm based on the WebMap. The goal is to locate suitable resources for any user request and to satisfy the informational need.

Establishing the intended WebMap requires preparation and processing of websites with the means of natural language processing (NLP). Related literature and approaches are introduced in Sect. 5.2. The fundamental concept of our new WebEngine is presented in Sect. 5.3.

5.2 Natural Language Processing Background

5.2.1 *Co-occurrence Graphs and TRC Extraction*

In the context of this work, co-occurrence graphs refer to graph databases containing nodes, that represent all words occurring in a given text corpus. Following [4], a co-occurrence graph is built by successively processing all documents of a text corpus and adding the words to the graph. Stopword removal and normalization (e.g., word stemming, or lemmatization) may be applied to merge grammatical variants of a word and remove irrelevant words. Nodes are then connected by edges whenever two corresponding words are co-occurrent, i.e., were observed in the same frame, e.g., a sentence, or a paragraph. The weight of an edge is based on the frequency with which the two words co-occurred. Its inverse can be interpreted as the distance. Figure 5.2 illustrates the process.

Determining a single representing term for a given document has been the subject of previous research. [5] presented the *text representing centroids* (TRC). In analogy to the geometric center of mass, it utilizes co-occurrence graphs to determine the graph node, i.e., term with a minimum average distance to all words of a given text. Other methods of term and keyword extraction, may also suit the later applications in the WebMap and should be reviewed. A comprehensive overview of methods was presented in [6].

5.2.2 *Distributed Global Graphs*

The size and connectivity of a co-occurrence graph increases, while documents are added and it will finally become a large, complex structure, where TRC extraction and other tasks become computationally expensive. A concept for the creation of a distributed global graph was introduced in [3], suggesting that parallelization and load distribution help to overcome these issues:

With a given set of participating peers $p_i \in P$, each peer p_i maintains a local co-occurrence graph L_i. First, the local graphs L_i of the initial peers are merged into a global graph G. The global graph is then divided into multiple sub-graphs, G_i, which are again distributed among all peers p_i. Subsequent changes in the local graphs are

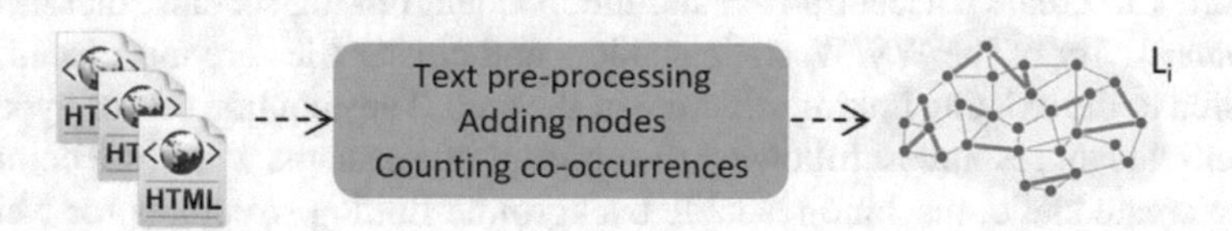

Fig. 5.2 A co-occurrence graph L_i is constructed by processing a given text corpus according to the described procedure

published to the global graph. Now, all operations, e.g., edge-weight calculation, or TRC extraction, can be performed on the global graph. It was shown, that the distributed global graph helps to increase efficiency, in terms of computational cost, load distribution, and parallelization.

While a local graph L_i is limited to a single peer's text corpus and allows TRC computation only in it's individual context, a mature global graph represents all peers' text corpora, and therefore, contains more terms and edges, than any local graph L_i. Besides increased efficiency, this allows to shift the computation of TRCs from an individual and specialized, to a more general level.

5.2.3 Documents and Connectivity of the WWW

In the scope of this paper, *documents* refer to web pages on the WWW. Web pages are addressable by their URLs and connectivity is achieved by hyperlinking among them. As the newly proposed WebEngine is based on routing between web pages, it needs to be ensured that they are sufficiently linked with other relevant and contextually related resources and that this linking follows a common syntax. Hence, connectivity needs to be increased and formalized. The question on how to create derive link recommendations with the help of co-occurrence graphs was addressed in previous publications:

The approach presented in [2] relies on a server-based P2P network, where each peer, i.e., web server, creates a local co-occurrence graph, based on the hosted documents. Each peer can translate a search query (or any other text) into its corresponding TRC and return all documents that are linked to the same TRC or a nearby term, i.e., node on the graph. Neighboring peers can be queried for matching documents, too.

In [7] a user-based approach of creating individual link recommendations was introduced. It is based on a P2P network in which similar users become neighbors and link recommendations are derived from neighbors' web-traverses. Here, every participating user needs to maintain a co-occurrence graph, store all traverses and share them with their neighbors. The approach, furthermore, incorporates neighbors' feedback and visitation frequencies to compute a ranking for the proposed links.

Both approaches compute on-demand link recommendations based on co-occurrence graphs. Again, this is computationally expensive and relies on the availability of the underlying P2P networks. Additionally, links are discarded immediately, and recommendations have to be recomputed each time.

In order to create a decentralized and independent linking scheme, that increases the connectivity of the WWW, static shadow and cluster files are introduced, as an extension to the existing linking structure of the web. They contain sets of hyperlinks to other relevant resources, following the concept of sitemaps, which are commonly used to create static, machine-readable background linking structures for websites. Likewise, shadow and cluster files need a predefined format and need to be addressable by a common naming scheme, to be universally accessible and processable. Their static nature aims to overcome the disadvantages of the aforementioned approaches, namely computational expense, dependency on the underlying P2P network, and lack of sustainability.

Given the idea of an extended, static linking on the WWW, the questions now are, how to utilize this to build a system, that

- establishes the WebMap, an extensive background linking scheme,
- allows routing from any starting point to any desired target,
- determines link recommendations based on a given input text or query.

5.3 The New Distributed WebEngine

5.3.1 Extending the Structure of the Web

5.3.1.1 Static Cluster Files and Cluster Graphs

Each document, i.e., web page, can be assigned to a TRC. As there exist documents sharing the same TRC, the idea is to create a cluster file containing hyperlinks to all documents that share the same TRC, i.e., $Term_i$. From all documents assigned to a cluster, a cluster co-occurrence graph L_i can be generated. This is analogous to the local co-occurrence graphs mentioned before, but builds upon all documents of the cluster $Term_i$, instead of the local documents. Next, a linking scheme among clusters can be achieved by enriching each node of the cluster graph L_i with a link to the corresponding term's cluster file $Term_j$. Figure 5.3 illustrates the relationship between documents, cluster files, and cluster graphs. The cluster graphs can later be utilized to build a global graph G, shared among all participating peers. Cluster files should be provided at standardized locations by the participating web servers for universal access.

5.3.1.2 Static Shadow Files

One of the goals of the proposed WebEngine is to design a static, inexpensive and far-reaching linking scheme within the underlying WWW. Therefore, in addition to the linking of cluster files and documents, each peer publishes static shadow files S_x for all documents or web pages D_x hosted. Each shadow file consists of four parts:

1. **Existing Links**: Links, that are already present on the web page
2. **Links to similar documents**: E.g., links to documents with the same TRC
3. **Link to associated cluster file**: Link to the document's TRC cluster file
4. **Links to other cluster files**: E.g., clusters of TF-IDF terms of the document

Figure 5.4 illustrates the composition of a shadow file S_x for a given document D_x and the linking of other resources. It also shows how the proposed linking scheme leads towards the intended WebMap and extends the existing linking structure of the

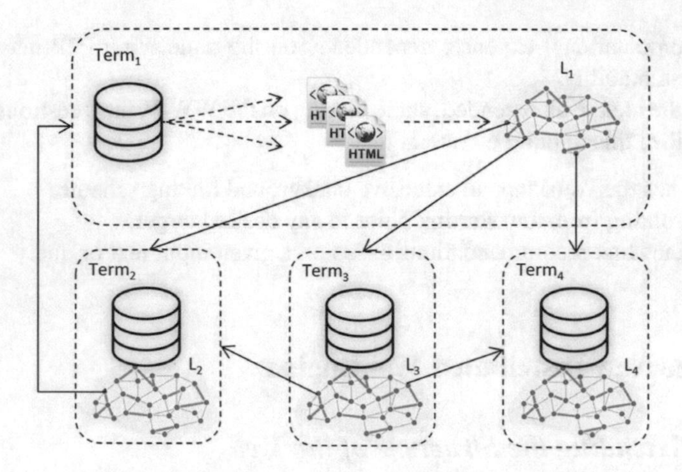

Fig. 5.3 Dashed arrows: Generating a cluster graph from documents listed in a cluster file. Solid arrows: Links from nodes of cluster graphs to corresponding cluster files

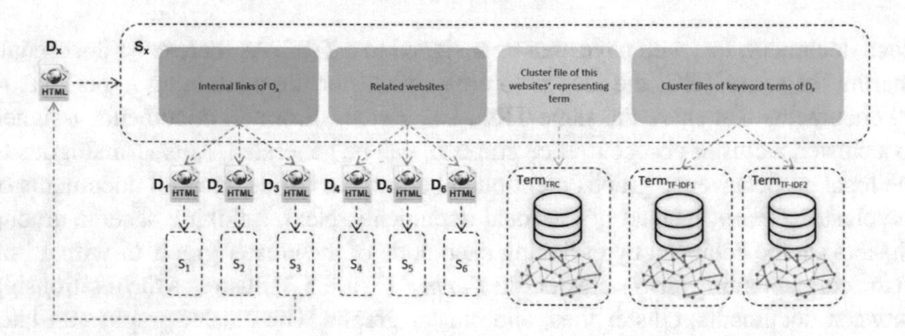

Fig. 5.4 Peers maintain shadow files S_x for all documents D_x. Shadow files contain links to other related documents and cluster files

web. Like cluster files, shadow files should be designed as simple text files, which should be periodically updated, to keep the cost of storage and computing low. Also, shadow files should be published with a standardized naming and location to ensure universal accessibility.

5.3.2 *Establishing the Decentralized WebEngine*

5.3.2.1 Global Graph and Distributed Clusters

The next step in setting up the WebEngine is to develop a system, that initializes and maintains the extended connectivity over time. Given the assumption, that multiple peers perform the same actions, i.e., create local clusters and cluster graphs, these can be merged into a global, distributed graph structure as discussed in Sect. 5.2.2 and proposed in [3]. Before the global graph is created, duplicate clusters that exist on the network need to be identified and merged. When this is accomplished, all terms and their associated cluster files and graphs, are represented uniquely on the network and can unambiguously be addressed by the nodes of the global graph.

Utilizing the distributed global graph as a point of reference, the clusters can now be distributed among all participating peers of the network. Each node of the global graph is, therefore, enriched with a set of data, allowing the routing towards related resources. It is composed of:

- A link to the term's corresponding cluster file
- Information of neighboring nodes hosted on remote peers, including peer and edge information

Figure 5.5 depicts the composition of the intended network and illustrates the linking between cluster files, cluster graphs, and the global graph. To build the intended global graph, several algorithms, and protocols need to be developed, or applied:

- Identifying duplicate clusters and merging them
- Establishing a global graph based on cluster graphs
- Distributing changes, to the entire network, e.g., cluster locations.

Since the authority to define a document's representing term, and thus the cluster-membership, lies with its owner, the peer managing the corresponding cluster is responsible to include a link to the document and update the respective cluster graph, as requested by the owner. Therefore, a mechanism of proving document ownership needs to be developed.

5.3.2.2 Searching for Cluster Files

As defined before, a document's static shadow file S_x contains links to the clusters $Term_i$ of a document's representing terms and keywords (e.g., TRC, TF-IDF). To establish this linking, the peer hosting the document, needs to locate the corresponding clusters on the network. The following approaches can be applied to achieve this:

1. **TheBrain approach**: This approach was introduced in [3]. It is based on the assumption that a path between any two nodes on any local co-occurrence graph is also represented in the global graph. Hence, checking the local graph and iteratively the neighboring peers' graphs for the requested term would finally deliver a path towards the desired node on the global graph. Translating TheBrain's approach into the scope of this work implies utilizing the local cluster graphs for this purpose.
2. **Crawler**: Independent from the underlying graph databases this approach attempts to crawl the static linking of shadow and cluster files. First, the crawler would periodically check existing links of a given document to discover at least one

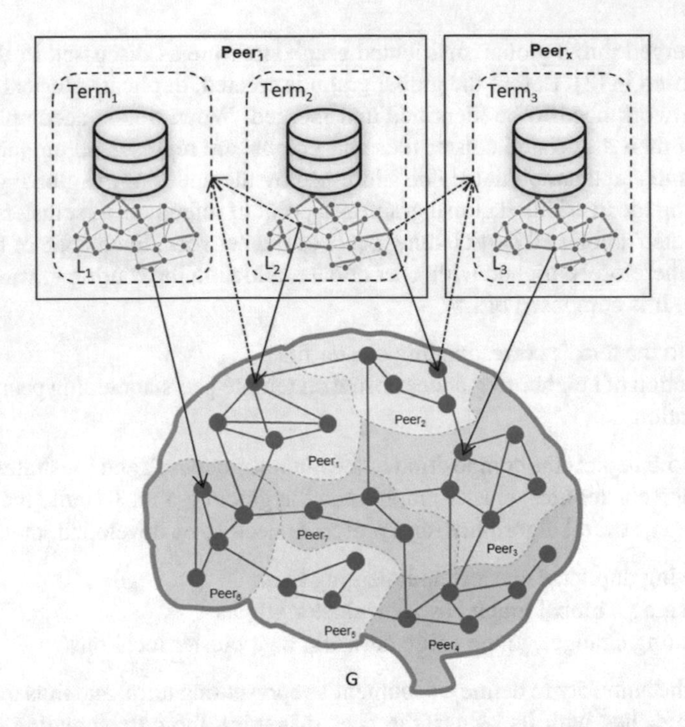

Fig. 5.5 Peers host cluster files and graphs, as well as a part of the global graph. Dashed arrows: Links from cluster graph nodes, and global graph nodes to the corresponding cluster. Solid arrows: Links from cluster graph nodes to global graph nodes. Solid edges on the global graph: Bi-directional links between nodes across multiple peers

document that maintains a shadow file. Second, the crawler extracts clusters from any detected shadow files. Third, clusters are crawled to identify more documents, which again contain links to other clusters. This is repeated until the crawler finally discovers the desired cluster.

3. **Random Walker**: Similar to the crawler, a random walker would locate clusters on the network, independently of the underlying graph databases. Unlike the crawler, a population of random walkers would also not consider the existing and extended static linking scheme as a pattern of movement, but randomly visits (participating) peers and queries their clusters.
4. **Creating a cluster**: In case that no cluster could be found, the peer creates a new cluster for the given term. This includes creating a cluster file and a cluster graph. If at any later point in time a duplicate cluster is found, the two clusters need to be merged as described in Sect. 5.3.2.1.

The focus of further research will be put on the second approaches, which relies on the extended static linking on the WWW, rather than the first approach which continues to depend on the underlying P2P network. Beyond the idea of crawling the static link structure, the crawling approach could, furthermore, be extended by a caching mechanism. This would, upon successfully locating a cluster, store the cluster's URL on all peers of the successful path, while any of the peers on the unsuccessful routes would keep a link towards the requesting peer. Later, requests for the same cluster, that pass any of these peers, can directly be routed to the desired resource.

5.3.3 Integrating Unmanaged Documents

In order to cover more documents than provided by the participating peers, each peer may utilize techniques such as crawling to identify documents on the WWW, whose hosts do not participate in the WebMap. Acting as a proxy, a peer may handle unmanaged documents, as if they were it's own. Accordingly, it can assign documents to clusters and maintain shadow files. Any other peer can validate if the providing peer is a proxy or the rightful owner by checking if the shadow file and the document are provided by the same host.

When the owning host of a document joins the WebMap it needs to take over the ownership from proxies by dissolving the previous cluster assignment and assigning the document to a (new) cluster. The peer also needs to create a shadow file or adopt it from the proxy. Finally, the global graph needs to be updated to reflect the changes.

The following aspects need to be considered when introducing document proxies, and accompanying processes need to be put in place:

- The choice of documents to proxy: A peer could either proxy all discovered unmanaged documents, or proxy only those that pass predefined criteria, e.g., matching a peer's contextual preferences.
- Multiple peers may act as a proxy for the same unmanaged document: A process to resolve disputes concerning cluster assignment and priority of peers needs to be defined.
- A reverse lookup algorithm to identify proxies for a given document needs to be developed.

5.4 Summary

5.4.1 Further Research

Several aspects of the introduced concept should be subject to follow-up research. First, experiments to create clusters, which represent an integral part of the linking mechanism, and the setup of a global graph as described in Sects. 5.3.1.1 and 5.3.2.1

should be conducted. Second, the necessity to mandate a single cluster assignment per document and a single cluster per term should be verified. Third, Mechanisms to resolve conflicts as stated in Sects. 5.3.2.1 and 5.3.3, i.e., merging duplicate clusters or proxies need to be elaborated.

Allowing multiple clusters per term may also create additional benefits: For example, they could account for different meanings of an ambiguous term. Furthermore, multiple clusters for the same, unambiguous word would create distributed pools of links. Determining a set of overlapping links in these clusters may be an approach to identify valuable resources. Hence, suitable mechanisms of splitting and merging clusters need to be elaborated on in further research.

Beyond open questions regarding the creation of clusters and construction of a static linking structure, several algorithms, and communication protocols necessary for the implementation of the intended WebMap were brought up in this paper and remain open tasks for further research and development.

5.4.2 Conclusion

This paper outlined the fundamental concept for the decentralized WebEngine 3.0, that aims to satisfy users' informational needs on the web: Namely, web-search, and link recommendation. In doing so, the introduced WebEngine does not rely on a central provider, or an active P2P network in on-demand processes compared to previous approaches, e.g., [2, 3]. Following the analogy of sitemaps shadow and cluster files constitute a static *WebMap*, that can be distributed across any number of participating peers. Routing and search algorithms can operate on this WebMap to locate desired resources.

References

1. Cockburn, A., Xu, G., Mckenzie, B.: Lost on the Web: An Introduction to Web Navigation Research (2001)
2. Kubek, M.: Concepts and Methods for a Librarian of the Web. Springer, Berlin (2019)
3. Simcharoen, S., Unger, H.: The Brain: WebEngine Version 2.0. In: The Autonomous Web, Springer (2022)
4. Srihari, R.K., Jin, W.: Graph-based text representation and knowledge discovery. In: Proceedings of the 2007 ACM Symposium on Applied Computing (2007)
5. Kubek, M., Unger, H.: Centroid terms as text representatives. In: Proceedings of the 2016 ACM Symposium on Document Engineering, pp 99–102 (2016)
6. Slobodan, Martinčić-Ipšić Sanda Beliga., Meštrović, Ana: An overview of graph-based keyword extraction methods and approaches. J. Inf. Org. Sci. **39**, 1–20 (2002)
7. Roßrucker, G.P.: Towards a new link recommendation indicator. In: The Autonomous Web, Springer (2022)

Chapter 6
Towards a New Link Recommendation Indicator

Georg P. Roßrucker

6.1 Introduction

Web-navigation has been a subject of scientific research for as long as the Internet exists. According to [1], the fundamental questions of a web-user regarding navigation are: "*Where am I?*", "*Where can I go?*", and most importantly, "*How do I get where I want to go?*". Today, various indicators of content growth on the Internet are observable. For example, the number of websites, i.e., domain names, is approaching two billion and has, over the last two decades, grown almost exponentially.[1] With the continuous increase of available content, the questions regarding web-navigation solidify.

In [2], the intentions of users were categorized by their needs: A *navigational need* is given when users know where they want to go but do not know the specific address. An *informational need* is given when users look for more information about a specific topic. Lastly, a *transactional need* is given when users seek specific web-services, such as online shopping, media consumption, or similar. These needs can partially be satisfied by the use of the fundamental web-navigation mechanisms discussed in [1], namely:

1. **Web-browser capabilities → navigational needs**: E.g., site visitation and re-visitation, navigating backward, forward, entering URLs, bookmarking
2. **Web-page design → navigational and informational needs**: The page design and static content links primarily aim to serve a user-friendly navigation within a specific website or domain
3. **Web-search engines → informational needs**: Deriving information resources from a search engine's index by submitting suitable search terms
4. **Specialized client and server-side tools → multi-purpose**: E.g., browser plugins, server extensions, and website features.

---

[1] http://www.internetlivestats.com/total-number-of-websites/.

H. Unger and M. Kubek, *The Autonomous Web*, Studies in Big Data 101,
https://doi.org/10.1007/978-3-030-90936-9_6

Concerning link recommendation and prediction of suitable subsequent pages for a given website, the informational need is to be satisfied, i.e., discovering topically related content and guiding the user on the shortest path towards the desired information [3]. Following the discussion in [1], *browser capabilities* do not satisfy informational needs beyond re-visiting websites. *Site-to-site hyperlinking* is only capable of providing new content, on the first visit, due to their static nature. Furthermore, concerns about document lifecycle, comprehensiveness, and bias of existing hyperlinks may arise. *Search engines and specialized tools* seem to be the most capable of satisfying the informational needs.

The initial design of most search engines was based on research regarding the structure of the web, aiming to detect and categorize valuable websites based on the underlying hyperlink structure of the Internet. The HITS algorithm [4] allows evaluating whether or not a website serves as a contextual match, i.e., as an *authority* for a given originating topic. PageRank, in contrast [5] computes a topic-neutral score for each website, which is iteratively propagated to hyperlinked websites, until an equilibrium is achieved. The disadvantages of both approaches are that they require high computing power, and the scores need to be updated regularly to reflect the continuously growing and altering structure of the web. Individual user preferences are also not reflected, resulting in search results, that are often broadly diversified and not tailored to the user's needs, making further filtering and review necessary [1]. An exemplary application is the iterative tracing of topics on the web introduced in [6].

Today, server-side recommender systems, based on social information filtering algorithms [7], e.g., collaborative filtering, are widely used in the field of e-commerce [8]. These algorithms are capable of providing recommendations like "*Other customers also bought X*" or "*You might also like Y*". This is usually achieved in the closed environment of a single domain or website by incorporating behavioral data, e.g., navigational sequences and categorization of users through clustering. Several of such sequence-aware recommender systems were introduced and analyzed in [9–11] and presented promising results, from which it can be inferred that it will be equally essential to consider navigational sequences and user similarity when designing a new link recommendation mechanism for the broader Internet.

The objective of this paper is to introduce just that: a new link recommendation indicator that works on the broader Internet, aiming to satisfy the informational needs of users and guiding them on the optimal path towards the desired information. The algorithm presented in Sect. 6.2 aims to identify and rate potential link candidates by bringing together sequence-based link discovery, contextual fitting, and consideration of the relevance of the proposed links for an individual user. Similar to the way product recommendation systems accomplish this task in closed environments, the approach presented here aims to extend the coverage of potential subsequent websites for a single user to the broader Internet, by decentralized data sharing across multiple similar users.

6.2 A New Link Recommendation Indicator

6.2.1 *The Link Recommendation Algorithm*

A new algorithm for website-to-website link recommendation will be proposed, which suggests subsequent websites for a given user and a given initial website. Out of scope are term or sentence based link recommendations, since this requires a more in-depth analysis of the content and methods of identifying suitable link-anchors. Also, the consideration of document lifecycles, i.e., actively crawling for new websites and observing changes, is not in the scope of this work. However, the algorithm implicitly considers document discovery. Potential use cases and applications are a browser-integrated recommendation of links, or a highlighting of valuable existing links in a given website.

The goal of the proposed link recommendation algorithm is to generate an ordered list of recommended subsequent websites (URLs), which helps to guide users via a minimal navigation path to the desired information. It is based on the assumption that users who are similar in terms of the context of previous site visitations will also have similar informational needs when researching the Internet. By querying a set of links from the stored web traverses of similar users, the number and quality of existing static and dynamic links can be narrowed down to a set of potentially suitable links. To rank these potential links, several rating criteria are introduced: A higher number of neighbors promoting a link is considered a positive indicator. Likewise, it is assumed that a higher user similarity indicates a higher favorability for a link. Furthermore, contextual similar pages are assumed to be more suitable than contextual-distant ones. Lastly, user's individual favorability with a given contextual is considered as a rating criterion that may also help to select and sort links ambiguation topics (preferred context of "mouse" for a biologist versus an IT-specialist). Table 6.1 lists all symbols necessary to compute the recommendation indicator for a set of potential subsequent websites.

By first completing a broad search on a user's neighbors, a set S of potential subsequent websites is generated and enriched with existing links from the given website. The algorithm then iterates through this set and calculates a link recommendation indicator ρ_x for each potential link candidate s_x. ρ_x takes a value between 0 and 1, where 1 denotes high recommendation, and 0 denotes no recommendation for a link. In doing so, the algorithm takes neighborhood similarity into account, as well as contextual relevance and user affection.

The following steps need to be executed, to obtain the desired output and are reflected by the outlined algorithm: For each potential subsequent website s_x compute

1. the neighborhood recommendation coefficient σ_x,
2. the contextual relevance coefficient γ_x, and
3. the user affection coefficient μ_x.

Table 6.1 Symbols used in the link recommendation algorithm

Symbol	Meaning
$u_j \in U_i$	Set of neighboring users of a given user i
$L_i(W, O)$	Local co-occurrence graph of user i
$F_i(K, E)$	Local web-traversing graph of user i
s_0	Initial website
$s_x \in S$	Set of potential subsequent websites to s_0
$\chi(L_i, s_x)$	The centroid of s_x, based on the local co-occurrence graph L_i
$C(L_i, \chi)$	The cluster to which a term, here centroid, belongs, based on a local co-occurrence graph L_i
$r(C(L_i, \chi))$	The individual rating of a given cluster
$\lambda_{i,j}$	Similarity coefficient of user i and user j, with $0 \leq \lambda_{i,j} \leq 1$
σ_x	Neighborhood recommendation coefficient for s_x, with $0 \leq \sigma_x \leq 1$: $\boldsymbol{\sigma_x} = \frac{1}{\lvert U_i \rvert} \sum_{j=1}^{\lvert U_i \rvert} \begin{cases} \lambda_{i,j}, & \text{if } F_j \text{ contains } s_x, \\ 0, & \text{otherwise.} \end{cases}$
γ_x	Contextual relevance coefficient for s_x, with $\gamma_x = 1 \vert 0$: $\boldsymbol{\gamma_x} = \begin{cases} 1, & \text{if } C(L_i, \chi(L_i, s_0)) = C(L_i, \chi(L_i, s_x)), \\ 0, & \text{otherwise.} \end{cases}$
μ_x	User affection coefficient, with $0 \leq \mu_x \leq 1$, derived from the user's individual cluster rating: $\boldsymbol{\mu_x} = r(C(L_i, \chi(L_i, s_x)))$
$\rho_x \in P$	Set of recommendation indicators for the potential subsequent websites S: $\boldsymbol{\rho_x} = \alpha * \boldsymbol{\sigma_x} + \beta * \boldsymbol{\gamma_x} + (1 - \alpha - \beta) * \boldsymbol{\mu_x}$
α, β	Arbitrary constants for weighting the coefficients in the computation of ρ_x, with $0 \leq \alpha, \beta \leq 1$ and $\alpha + \beta \leq 1$

These are then joined in the formulation of the link recommendation indicator ρ_x. ρ_x is added to the set P, which can finally be sorted descendingly, to obtain the highest-ranking link recommendations.

Inputs to the algorithm:

1. User u_0, maintaining a local co-occurrence graph L_0 and a local web-traversing graph F_0
2. A set of neighboring users $u_j \in U_i$ and their respective graphs L_j and F_j
3. Similarity coefficients $\lambda_{i,j}$ for user $i = 0$ and all neighboring user j, with $0 \leq \lambda_{i,j} \leq 1$
4. An initial website s_0

Initialization:

1. Create a set S of potential subsequent websites to s_0 with $s_x \in S$:
   a. Collect all nodes subsequent to s_0 from all neighboring users' u_j traversing graphs
   b. Collect all existing links from s_0
2. Define the constants α and β for appropriate weighting of the coefficients

Iteration:

For each $s_x \in S$:

1. Compute the neighborhood recommendation coefficient $\boldsymbol{\sigma_x}$ of s_x:

$$\boldsymbol{\sigma_x} = \frac{1}{|U_i|} \sum_{j=1}^{|U_i|} \begin{cases} \lambda_{i,j}, & \text{if } F_j \text{ contains } s_x, \\ 0, & \text{otherwise.} \end{cases}$$

2. Determine the contextual relevance coefficient $\boldsymbol{\gamma_x}$ of s_x:

$$\boldsymbol{\gamma_x} = \begin{cases} 1, & \text{if } C(L_i, \chi(L_i, s_0)) = C(L_i, \chi(L_i, s_x)), \\ 0, & \text{otherwise.} \end{cases}$$

3. Derive the user affection coefficient $\boldsymbol{\mu_x}$ of s_x:

$$\boldsymbol{\mu_x} = r(C(L_i, \chi(L_i, s_x)))$$

4. Compute the recommendation indicator $\boldsymbol{\rho_x}$:

$$\boldsymbol{\rho_x} = \alpha * \boldsymbol{\sigma_x} + \beta * \boldsymbol{\gamma_x} + (1 - \alpha - \beta) * \boldsymbol{\mu_x}$$

5. Add ρ_x to the set of resulting recommendation indicators $\boldsymbol{P}$.

Finalization and outputs of the algorithm:

1. Sort P descendingly, to obtain the highest-ranking link recommendations
2. Output: P

As a prerequisite, every user i maintains a local web-traversing graph F_i with $F_i(K, E)$, where the nodes K represent visited websites, and the directed edges E refer to the chronological sequence in which the user visited them. Sequences within this graph are paths of pages, subsequently visited during a time-limited web-session, which hence represent a subset of the real web-graph enriched with individual patterns of web-navigation. They are terminated and restarted for each new session, as well as when a subsequent page belongs to a different contextual cluster.

Additionally, every user i maintains a local co-occurrence graph L_i with $L_i(W, O)$, where the nodes W represent words occurring in websites, visited by the user, and the edges O denote co-occurrences and their number. In [12] was shown that co-occurrence graphs converge to an equilibrium with a rising number of documents processed. Therefore, they may be handled as stable graphs in the long-run.

It is assumed that *similar* users are connected in a P2P network and are permitted to share specific data:

1. A comparable metric, derived from the local co-occurrence graph, to determine the similarity coefficient $\lambda_{i,j}$ for two connected users, and
2. Sequences of the web-traversing graph F_i for a given website (s_0).

The similarity coefficients $\lambda_{i,j}$ for a user i and his neighboring users $u_j \in U_i$, are computed by, applying similarity metrics on the local co-occurrence graphs F_i and F_j [13]. Besides the computation of neighborhood recommendation coefficients σ_x, they also support the formation of the underlying P2P network, such that only similar users are being connected.

For each $s_x \in S$, the neighborhood recommendation coefficient σ_x is computed by adding up the similarity coefficients of all neighbors promoting s_x. The sum of similarity coefficients is divided by the total number of neighbors $|U_i|$. As a result, the neighborhood recommendation coefficient ranges from 0 to 1 and increases if more neighbors promote a website s_x.

The algorithm requires that the co-occurrence graph L_i is enriched with cluster and rating information. In [14], a suitable approach for co-occurrence-based graph clustering was introduced, which can be applied here. Each s_x of the collection S can be assigned to a cluster, by first determining its centroid term $\chi(s_x)$ [15] and then deriving its cluster-membership $C(L_i, \chi(s_x))$. The contextual relevance coefficient γ_x is defined as a binary value, indicating whether or not s_0 and s_x belong to the same contextual cluster.

The cluster rating is a representative rating, derived from individual ratings and feedback for websites that were assigned to a respective cluster. For each s_x, the user affection coefficient μ_x can be adopted from its associated cluster rating: $r(C(L_i, \chi(L_i, s_x)))$.

Finally, the recommendation indicator ρ_x for each s_x of S is computed by summing up the three coefficients into a final score. α and β allow the weighting of the three coefficients, where the ideal balance needs to be investigated in further experiments. The weighted neighborhood recommendation coefficient $\alpha * \sigma_x$ will default to 0 if the neighborhood does not promote s_x. Promoted links will, therefore, have a better ranking than otherwise equal but unpromoted links, i.e., existing links only collected from s_0.

It is crucial to bear in mind that the recommendation indicator ρ_x is not comparable across multiple users: The neighborhood recommendation coefficient σ_x depends on the number of neighbors and their similarity, which differs across users, and the ratings and cluster memberships are computed based on individual preferences.

6.2.2 Data Flow

Figure 6.1 illustrates the data flow of the introduced link recommendation algorithm. The different background colors outline functional areas, to which the components belong to. The connecting arrows indicate the flow of data, and the annotations give a brief description of the content, where required.

As illustrated, a user interacts only with the browser of the local PC, i.e., pc_1. The components of pc_1 are depicted as an example for all participating PCs. Each PC hosts a co-occurrence graph and a web-traversing graph, which are fed with content (documents) and user navigation-flows by the browser. The profile-worker enriches the co-occurrence graph with additional information, i.e., contextual clusters and transforming user behavior into cluster ratings. Similarly, the traversing-worker manages information on the web-traversing graph, i.e., splitting sequences by session or context. The web-traversing graph provides link candidates to neighboring nodes and the local link-collector via the peer-to-peer (P2P) node.

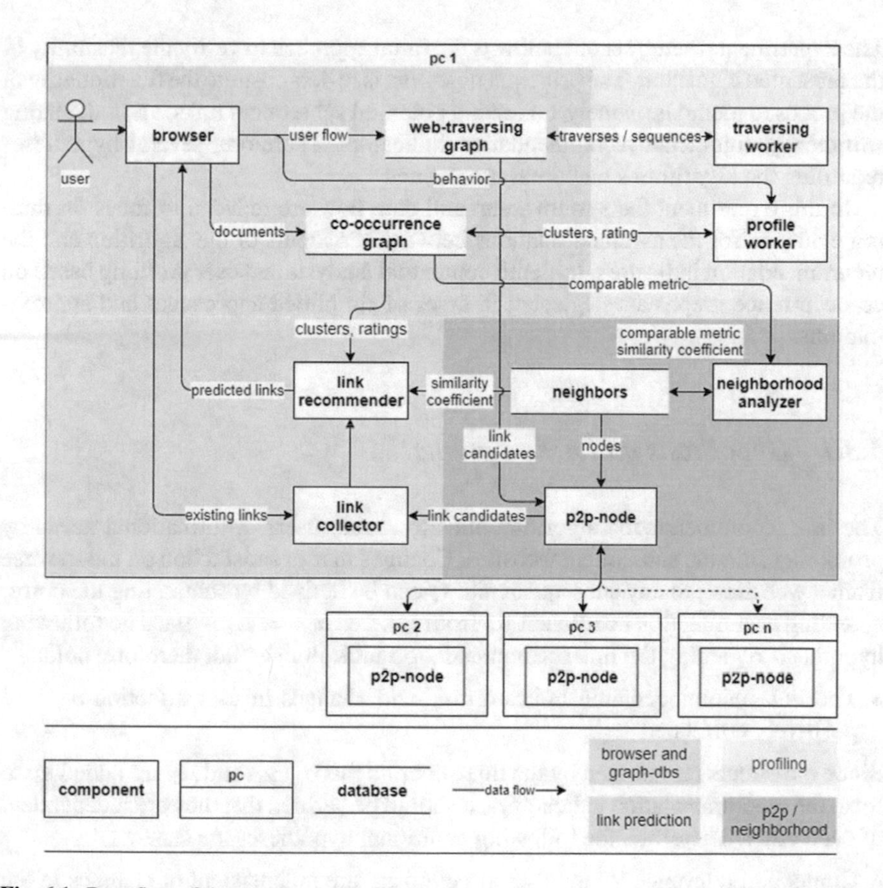

Fig. 6.1 Data flow of the presented algorithm

The computer pc_1 is part of a P2P network in which similar nodes are connected to each other. The connection is established via a locally running P2P node. The neighbors are negotiated by analyzing comparable metrics of the local co-occurrence graph with those of other members of the P2P network. Similarity and neighborhood information is stored in a local database and provided to the link-recommender unit.

The local link-collector takes the current website s_0 as an input from the browser. Potential link candidates are collected from the local web-traversing graph, existing links in s_0, and neighboring web-traversing graphs. The links are passed on to the link-recommender unit, which computes the recommendation indicator for all links based on inputs of the co-occurrence graph and neighborhood information. The resulting link-recommendation-list is sorted descendingly and finally returned to the browser for visualization.

6.3 Experimental Setup and Execution

The experimental setup set out below is the inital approach to verify the feasibility of the presented algorithm. The principal objective is to demonstrate the functionality of the proposed methods, namely, collecting potential subsequent links, and calculating sufficiently significant recommendation indicators. Therefore, several hypotheses regarding the algorithm's outcomes are defined.

In this experiment the system setup and data flow are reduced to focus on finding evidence for the assumed relations between the inputs of the algorithm and the recommendation indicators. In-depth contextual analysis and user profiling based on co-occurrence graphs are neglected in favor of simplified approaches and approximations.

6.3.1 Hypothesis and Assumptions

The link recommendation algorithm aims to satisfy users' informational needs by promoting suitable subsequent websites. Changes in user satisfaction on the traverse from a website s_0 to any subsequent site s_x can be derived by subtracting the corresponding user affection coefficient μ_0 from μ_x, i.e., $a_{0,x} = \mu_x - \mu_0$. The following hypothesis regarding the link recommendation indicator should, therefore, hold:

- The link recommendation indicators ρ_x and changes in user affection $a_{0,x}$ are positively correlated

Since the effects represented by the three coefficients σ_x, γ_x, and μ_x are added up to form the recommendation indicator ρ_x it should be verified that they are independent of each other. Therefore, the following additional hypotheses are stated:

- Contextual relevance γ_x and user affection μ_x are independent of changes in the neighborhood recommendation σ_x

- Neighborhood recommendation σ_x and user affection μ_x are independent of changes in the contextual relevance γ_x
- Neighborhood recommendation σ_x and contextual relevance γ_x are independent of changes in the user affection μ_x

6.3.2 Collecting and Pre-Processing Experimental Data

Following the 5th Workshop on Complex Structures 2020,[2] a plugin for the web-browser Google Chrome was created and published,[3] which aims to collect data for the generation of individual web-traversing graphs. The plugin tracks the web-navigation flow of a user and submits pseudonymized data to a server. The record of each traverse contains the following data:

- $from$—the origin of the user, i.e., last visited URL,
- to—the target of the traverse, i.e., URL of the next website,
- $timestamp$—a UNIX timestamp, and
- $httpstatuscode$—the HTTP status code returned by the target's web-server.

The pseudonymized data were used to generate sequences by concatenating matching to- and $from$-attributes of consecutive records, also allowing for branching of sequences. Each sequence is identified by a unique id and terminated if:

- the last $httpstatuscode$ is greater or equal 400,
- the break-duration, i.e., the difference between two traverses' $timestamp$ -attributes is greater than a predefined maximum break-duration, or
- no further records can be attached (e.g., start of a new session).

Sequences were entered into a web-traversing graph by creating nodes for each $from$ and to attribute. To avoid duplicates, commonly used query attributes such as *sid* or *session* and hash attributes were removed, before an URL was added to the graph. Edges between nodes were established, representing the direction of traverses and were enriched with a sequence-tag, containing an array of associated sequences and timestamps.

6.3.3 Adjusting the Algorithm

As this experimental setup intends to find evidence for the above-stated hypotheses without presuming the fully developed system setup and data flow introduced in Sect. 6.2.2, the algorithm needs to be adjusted to allow a similar interpretation of

[2] https://www.fernuni-hagen.de/kn/complex-structures/.

[3] https://chrome.google.com/webstore/detail/surfalizer/ifedcfnpmoioajmcmgnnnheafjlheacm.

coefficients without the presence of user-specific co-occurrence graphs and a large number of participating users.

Adjusting σ_x:
Following the algorithm, σ_x is based on co-occurrence inferred similarities of users promoting a website s_x. Without co-occurrence graphs and a variety of users, the effect of user similarity on user affection is tested for arbitrarily chosen values between 0 and 1.

Adjusting γ_x:
As a measure of contextual relevance, the cosine similarity of s_0 and s_x will be determined and replace the same-cluster assessment of the co-occurrence based approach described in the algorithm. The cosine similarity ranges from 0: no similarity to 1: high similarity.

Adjusting μ_x:
Instead of deriving the cluster rating from the local co-occurrence graph, the user affection is manually sampled for each website. Possible ratings are 0: no rating, 1: not relevant, 2: relevant, or 3: highly relevant. The rating is then divided by 3 to obtain a user affection coefficient μ_x between 0 and 1.

6.4 Results and Discussion

Due to an insufficient number of participants, the experimental dataset only consists of the author's web history between May 2nd, 2020, and June 2nd, 2020. To allow the application of the introduced link recommendation indicator, three fictitious users, 0, 1, and 2, were introduced, where users 1 and 2 are providers who share their sequence data with the relying user 0.

6.4.1 *Web-Traversing Graph*

From the recorded dataset, a web-traversing graph was derived. Some general statistics of the graph are presented in Table 6.2. Since traverses have reoccurred over time, their number is greater than the number of edges. However, all occurrences were recorded and attached to the edges as an array of sequence ids and timestamps.

Table 6.2 Key figures of the web-traversing graph

Recorded traverses	Number of nodes	Number of edges
6700	2323	3505

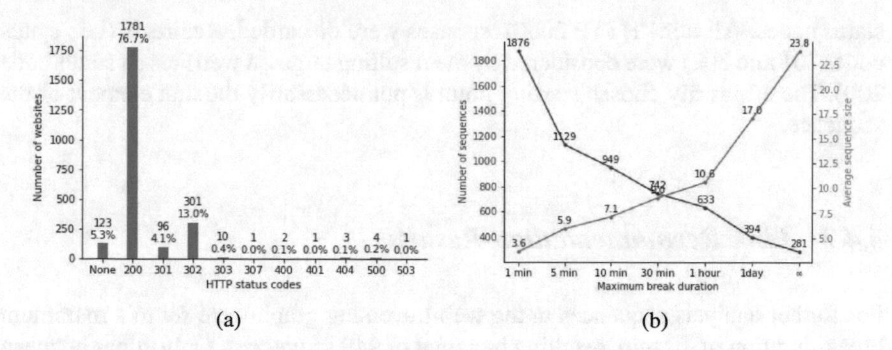

Fig. 6.2 **a** Number of websites per HTTP status code of recorded traverses, **b** Maximum session break-duration and resulting number of sequences and average sequence size

Table 6.3 An example of a sub-sequence for a given starting point

Sub-sequence starting at https://www.marina-guide.de/
https://www.marina-guide.de/
→ https://www.marina-guide.de/erweiterte-suche/ (200)
→ http://marina-guide.de/marinas (200)
→ https://www.marina-guide.de/marina/marina-yacht-park-[...] (200)
→ https://www.marina-guide.de/marina/milford-marina/ (200)
→ https://www.marina-guide.de/harbour/milford-marina/ (200)
→ https://www.marina-guide.de/advanced-search/ (200)
→ https://www.marina-guide.de/erweiterte-suche/?search[...] (200)
→ https://www.marina-guide.de/marina/marina-recke/ (200)
→ https://www.google.com/search?q=googlemaps+api&oq=[...] (200)
→ https://www.marina-guide.de/marina/marina-yacht[...] (200)

Each website, i.e., node of the graph, was enriched with the returned HTTP status code.[4] Figure 6.2a presents the number of websites per HTTP status code. The websites, for which the recording of the HTTP status code failed, are listed as *None*.

Based on the traverses and timestamp-properties of edges, sequences were generated as defined in Sect. 6.2. The number of sequences and their sizes vary with different presets for permitted break-durations of sessions. Figure 6.2b depicts the resulting numbers for different values of maximum break-durations, i.e., 1, 5, 10, 30 min, 1 hour, 1 day, and unlimited.

Given an unlimited maximum break duration, the sequence in Table 6.3 was extracted from the web-traversing graph for demonstration purposes. Here https://www.marina-guide.de/ was used as a starting point. Consequently, there exists only one sequence containing all outgoing traverses from this starting point. The arrows depict the direction of each traverse. The numbers in brackets reflect the HTTP

[4] https://tools.ietf.org/html/rfc2616#section-10.

status codes. All non-'HTTP 200' responses were discarded. Redirects (i.e., status codes 301 and 302) were considered by the resulting targeted websites (if status code 200). The arbitrarily chosen starting point is not necessarily the first element of the sequence.

6.4.2 Link Recommendation Results

For further analysis sequences in the web-traversing graph were set to a maximum break-duration of 10 min, resulting in a total of 949 sequences. Only edges between nodes with HTTP status codes equal to 200 were considered, reducing the number of edges to 2363. The sequences were then alternately assigned to the fictitious users 1 and 2, depending on their sequence-id. All sequences with even ids were assigned to user 1, those with odd ids to user 2.

The three coefficients σ_x, γ_x, and μ_x were computed for each traverse of the experimental dataset as defined in Sect. 6.3.3. Since the sequences of the fictitious users 1 and 2 emerged from the same dataset, the similarity coefficients $\lambda_{0,1}$ and $\lambda_{0,2}$ could not be calculated as a result of any similarity metric. To demonstrate the effect of different values on the resulting link recommendation indicator, the similarity coefficients for users 0 and 1, as well as 0 and 2, were arbitrarily chosen and differ slightly:

$$\lambda_{0,1} = 0.8\ ;\ \lambda_{0,2} = 0.6$$

Finally, the link recommendation indicator ρ_x was calculated for the target website of each recorded traverse. The weighting of the three coefficients was disregarded, i.e., giving them equal weights in the computation of ρ_x:

$$\alpha = 0.33\ ;\ \beta = 0.33\ ;\ (1 - \alpha - \beta) = 0.34$$

The scatter plots in Figures 6.3 and 6.4 illustrate the values of the calculated coefficients and recommendation indicators of each traverse, helping to verify the four hypotheses.

Figure 6.3 contains six diagrams displaying the values of the link recommendation coefficients. For each horizontal pair of plots, the list of traverses was sorted for one coefficient and the values of the remaining two coefficients were drawn up against it. Plots (a) and (b) are sorted by the neighborhood recommendation coefficient σ_x. The values of contextual relevance γ_x and user affection μ_x are drawn up against it, illustrating that for each level of neighborhood recommendation both coefficients range almost evenly from very high to very low values. This suggests that they are both unrelated to the neighborhood recommendation coefficient. Similarly, the plots (c) and (d) illustrate that neighborhood recommendation σ_x and user affection μ_x are not related to a changing level of contextual relevance γ_x, i.e. their distribution behaves unchanged, regardless of the level of contextual relevance. Lastly, the plots (e) and (f) display the values of contextual relevance γ_x and neighborhood recommendation

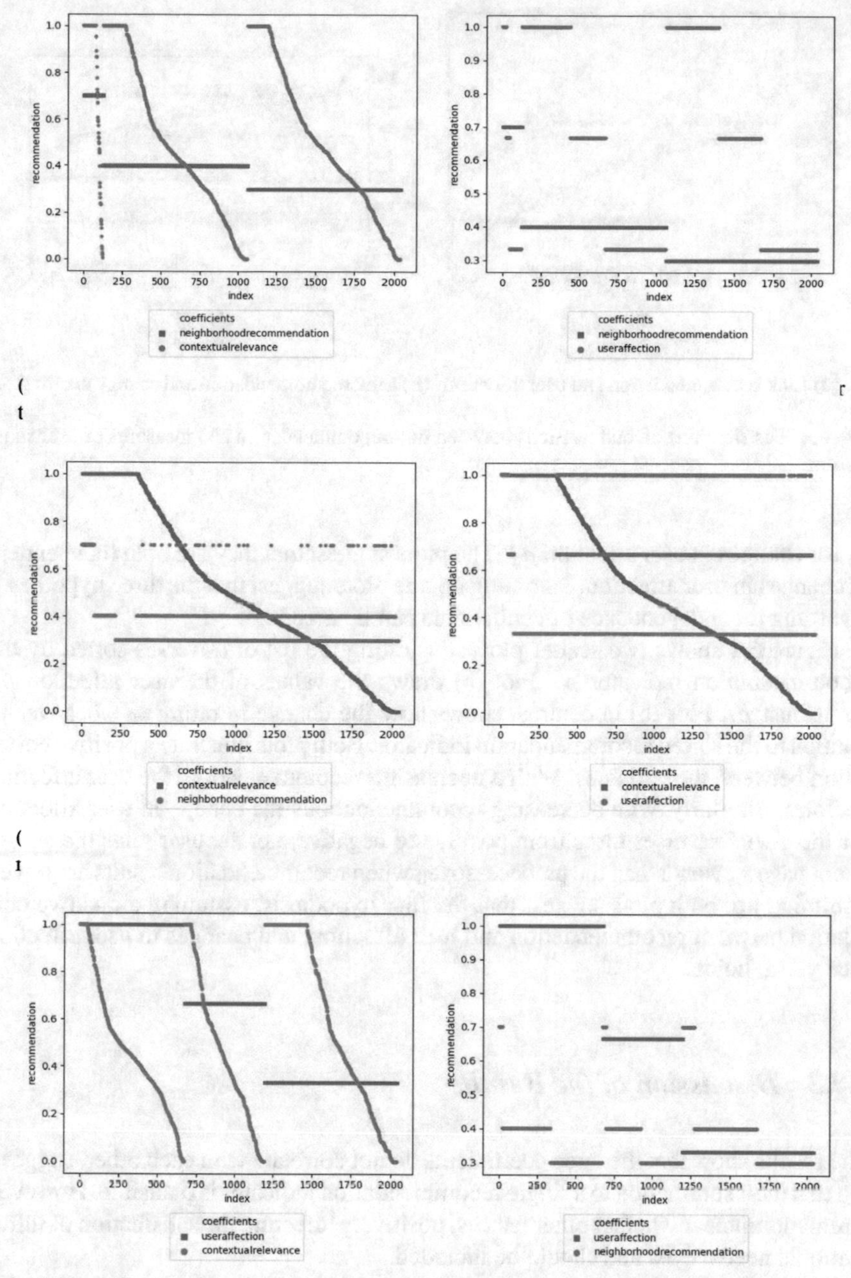

(e) User affection and contextual relevance (f) User affection and neighborhood recommendation

Fig. 6.3 Distributions of, and relations between link recommendation coefficients

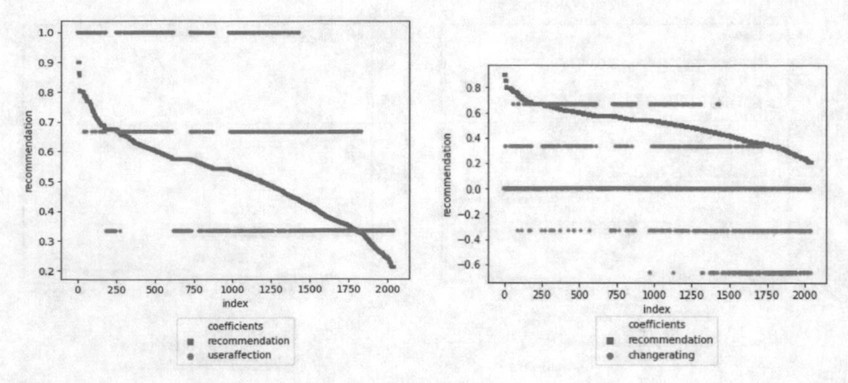

(a) Link recommendation and user affection (b) Link recommendation and change in rating

Fig. 6.4 Distributions of, and relations between link recommendation and measures of user satisfaction

σ_x for changes in user affection μ_x. The plots suggest that they are both independent of changes in user affection. Summing up, the plots suggest that the three hypotheses regarding the independence of coefficients can be accepted.

Figure 6.4 shows two scatter plots illustrating the list of traverses sorted by the recommendation indicator ρ_x. Plot (a) draws the values of the user affection μ_x up against ρ_x. Plot (b) in contrast shows how the change in rating $a_{0,x}$ behaves in relation to the sorted recommendation indicator. Both plots illustrate a positive correlation between the variables. With a decreasing recommendation, the user affection declines. Similarly, with decreasing recommendations the change in user affection for the given traverses turns from positive to negative, i.e. the user rates the subsequent page s_x worse than the predecessor s_0 when recommendation results are lower. Summing up, both plots suggest that the first hypothesis, assuming a positive correlation between recommendation and user affection, and changes in user affection vice versa, holds.

6.4.3 Discussion of the Results

The results show that the three coefficients do not correlate with each other, suggesting that their summation to a single recommendation indicator is plausible. However, it remains unclear whether other factors, positively affecting the satisfaction of informational needs, exist and should be included.

Due to the poor dataset, the assumption, that the three coefficients have positive effects on the satisfaction of informational needs, remains unproven. Apart from the question of whether or not the three coefficients affect the satisfaction of informational needs, it remains also unclear how strong this effect is. Depending on this, the weightings, α, and β would have to be chosen.

It was shown, that a higher level of user satisfaction exists for those websites, that received a higher recommendation score, suggesting that the algorithm is capable of generating desired results. Another interesting finding is, that high recommendations seem to go along with an increases of the rating of the websites. The results showed that above a certain recommendation level (roughly 0.5) the changes in rating, i.e. user satisfaction with the recommended links, are mainly positive.

Despite the promising results, the proposed algorithm limits the choice of link candidates to the first subsequent links of neighbors' traversing-graphs. Yet, in Sect. 6.4.1, it was shown that sequences can contain multiple subsequent URLs, and it is often the case that the informational need is satisfied only after multiple traverses. Better link candidates might be identified by evaluating the full subsequent chains on the traversing-graph.

6.5 Summary

It was shown that with a given set of prerequisites, the introduced link recommendation algorithm is capable of computing link recommendations for a given website and user. Following the setup outlined in Sect. 6.2.2, the algorithm would also be capable of performing this task in real-time.

The web-traversing graph and its application were the primary subjects of the initial experiment. Herein, the graph successfully served as a source for potential subsequent links, built upon real browsing data of a single user. Unfortunately, the calculation of coefficients was based on supportive functions and approximations and not on real co-occurrence graphs, as originally proposed in Sect. 6.2. This should be repeated with a suitable dataset, to verify the results of this work.

It can be concluded, that with the help of the proposed link recommendation algorithm, the objective of satisfying users' informational needs is achievable. The results furthermore suggest, that there exists a recommendation threshold, above which the change in rating is found to be mainly positive. Identifying this threshold and other conditions would help to determine significantly valuable link candidates. This should be elaborated on in further research.

References

1. Xu, G., Cockburn, A., Mckenzie, B.: Lost on the Web: An Introduction to Web Navigation Research (2001)
2. Broder, A.: A taxonomy of web search. SIGIR Forum **36**(2), 3–10 (2002)
3. Gwizdka, J., Spence, I.: Predicting outcomes of web navigation. In: Proceedings of the 14th international conference on World Wide Web, WWW, pp. 892–893. Chiba, Japan (2005)
4. Kleinberg, J.: Authoritative sources in a hyperlinked environment. J. ACM **46**(5), 604–632 (1999)

5. Page, L., Brin, S., Motwani, R., Winograd, T.: The PageRank Citation Ranking: Bringing Order to the Web (1999)
6. Kubek, M.M.: Interaktive Anwendungen kontextbasierter Suchverfahren, VDI Verlag, Düsseldorf (2014)
7. Shardanand, U., Maes, P.: Social Information Filtering: Algorithms for Automating 'Word of Mouth'. In: Proceedings of the Conference on Human Factors in Computing Systems, pp. 210–217. ACM Press (1995)
8. Schafer, J.B., Konstan, J., Riedl, J.: Recommender Systems in E-Commerce. In: Proceedings of ACM E-Commerce (1999)
9. Quadrana, M., Cremonesi, P., Jannach, D.: Sequence-aware recommender systems. ACM Comput. Surv. 1, **1**, Article 1 (2018)
10. Mishra, R., Kumar, P., Bhaske, B.: A web recommendation system considering sequential information. Decis. Support Syst. **75**, 1–10 (2015)
11. Forsati, R., Moayedikia, A., Shamsfard, M.: An effective Web page recommender using binary data clustering. Inf. Retrieval J. **18**(3), 167–214 (2015)
12. Unger, H., Kubek, M.M.: Theory and Application of Text-Representing Centroids. VDI (2019)
13. Koutra, D., Ramdas, A., Xiang, J.: Algorithms for Graph Similarity and Subgraph Matching 201
14. Simcharoen, S., Unger, H.: Dynamic clustering for segregation of co-occurrence graphs. In: Autonomous System Conference, pp. 53–71. Majorca, Spain (2019)
15. Kubek, M., Unger, H.: Centroid terms as text representatives. In: Proceedings of the 2016 ACM Symposium on Document Engineering, pp. 99–102 (2016)

Chapter 7
A Survey on Innovative Graph-Based Clustering Algorithms

Mark Hloch, Mario Kubek, and Herwig Unger

7.1 Introduction and Motivation

Grouping, e.g. images, terms or documents, into groups of similar objects is referred as the process of clustering. In contrast to supervised classification, where the algorithm has been trained how to map its input to an according output, clustering is performed unsupervised and therefore mainly relies on the provided input data to find the best grouping of objects based on this inherent information. Due to the great variety of input data to be clustered, the according fields of application for such algorithms are manifold, thus there are dozens of clustering algorithms been developed over the last decades [1, 2].

Classical, typically vector-based algorithms, such as the kmeans [3], kmeans++ [4] or k-NN algorithm [5] require to choose a hyperparameter, as the suggested number of expected output clusters, a priori. In cases where the data to be clustered changes over time, such as in search-engines, this preliminary requirement forces the user to estimate beforehand, what number of result clusters are expected. This approch therefore softens the idea of an unsupervised algorithm providing the best possible result fully automatically, without user intervention. Moreover, this estimation has to be performed any time the input data is changed significantly. This clearly softens the general idea of having a fully unsupervised algorithm that perform automatic clustering providing the best possible results. In addition, actual standard algorithms mentioned above, but also newer graph based algorithms, e.g. chinese whispers [6] expose another weakness as they are not designed to adapt to growing or shrinking corpora: they typically require a full set of documents beforehand. With the focus of the authors preliminary work on the concept of the librarian of the web [7], the mentioned disadvantages also prevent a better usability to the user: instead of having a long list of search results on a user's query the knowledge that lies within the available document corpus, can be presented much more easily [8] using clusters of best matching documents.

With the work of [9, 10] it is shown that the use of co-occurrence graphs is very useful for graph-based concepts on which novel clustering technques can be built on.

H. Unger and M. Kubek, *The Autonomous Web*, Studies in Big Data 101,
https://doi.org/10.1007/978-3-030-90936-9_7

Recent work on graph based clustering algorithms [11, 12] provide novel approaches in the field of clustering. This paper gives a comparison of these graph-based algorithms and shows their benefits over classical approaches, as well as requirements for further optimization.

7.2 Classical Clustering Algorithms

7.2.1 K-Means

The k-means [3] algorithm is one of the most classical hard-clustering algorithms used. Due to its straightforwardness it can be easily implemented and applied in manifold use-cases, e.g. customer segmentation [13] or cyber-profiling criminals [14] as well as classical document clustering. From performance point of view the kmeans algorithm performs, due its linearly proportional complexity to the size of datasets, even on larger datasets efficiently.

The underlying idea of the k-means algorithm is to create k distinct partitions by setting cluster centers that are updated in an iterative manner. As shown in the pseudocode of the algorithm given below the algorithm mainly consists of three steps. Firstly the user has to define manually the value k , which determines how many clusters should be a result of k-means. Secondly, a loop of two repeating steps

- of assigning the data set points to one of the clusters with lowest distance to the clusters centroid and A Survey on Innovative Graph-based Clustering Algorithms
- the calculation of a new centroid for each cluster.

The clustering result itself can be influenced by mainly three parameters that can lead into disadvantages:

I. choice of distance measure,
II. chosing the initial centroids and
III. estimating the hyperparameter k.

The distance measure is involved in the computation of the clusters centroids and therefore has an impact on on the clustering result. Aggarwal and Reddy [15] gives an overview on possible distance measures, such as the manhatten or cosine similarty, applicable. In general, the most common distance measure used is the Euclidean distance.

K-means algorithm

Input: A set of input points X and number of k clusters to be created
Output: A set of k clusters

a. **Initialisation**: let k random input data points $x \in X$ be the initial clusters C.
b. **Loop**: until the cluster centroids do not change significantly or other convergence criteria are met

- Assign each sample $x_i \in X$ to its nearest centroid $c_i \in C$ by using the least squared Euclidean distance $\arg\min d(c_i, x)^2, c_i \in C$.
- Recompute the (new) centroid c_i of each cluster, which is defined as the mean of the according data points x in cluster i.

In the work of [16] it is shown that clustering result and performance of the k-means algorithm depends on the initialisation of the cluster centers. The initialisation of the cluster centers therefore has been suspect of research over the years and lead to several approaches e.g. Forgy [17], Hartigan [18] or [19]. In general the most commonly used initialisation is, to use a number of k randomly chosen initial clusters as originally suggested in the work of MacQueen [3].

The requirement of manually estimating the number of clusters a priori is a great limitation for the k-means algorithm to be applied in use-cases where the amount of data to be clustered is changing over time. From the practical point of view this would require the user to estimate a good k value before actual clustering can be performed. Especially in cases where the input data are very large or growing over time the manual selection of k is therefore not applicable. Approaches, e.g. the Silhouette Coefficient [20] or Calinski-Harabasz Index [21] can be used to resolve this problem by suggesting the k-value automatically.

7.2.2 *K-Means++*

A very commonly used extention to the classical k-means algorithm is k-means++[4], which mainly aims to make an improvement on choosing the cluster centroids more accurately during initialisation.

As outlined below k-means++ in contrast to k-means initially uses randomly chosen points as cluster centers and then subsequently uses a weighted probability score to improve the finding of cluster centers over time.

K-means++ algorithm

Input: A set of input points X and number of k clusters to be created
Output: A set of k clusters

a. Choose a single random cluster center c_i from the set of data points D
b. **Loop:** until k cluster centers found

   Chose the next cluster center as the most distant to the previous one by using a weighted probability score

c. Perform standard k-means algorithm

Due to the optimization of the cluster initialisation the clustering results are improved. As the k-means++ algorithm only optimizes the initialisation part of the k-means algorithm, the disadvantage of manually estimating the hyperparameter k still remains.

7.2.3 Mini-Batch k-Means

Mini-batch k-means or fast k-means [22] is a modified version of the k-means algorithm that aims to improve the scalability and performance for large datasets such as web applications.

The mini-batch k-means algorithm as outlined below takes small random batches of data and assigns each of the sample points to a centroid. In a second step the cluster centroid is then updated based on the streaming average of all of the previous samples assigned to that centroid.

Mini-batch k-means algorithm

Input: A set of input datapoints X, k-value and number of iterations j
Output: A set of k clusters
Initialisation: get k random datapoints $x \in X$ as initial clusters centers c_i

a. **Loop:** until j iterations performed

   i Get batch dataset M of size b; let $x_m \in M$ be random data taken from X
   ii **Loop:** over each x_m with $1 \leq m \leq b$

   – Cache the cluster center μ_{c_i} with shortest distance to x_m

   iii. **Loop:** over each x_m with $1 \leq m \leq b$

   – Get the cached center μ_{c_i} for x_m
   – Update the sample number $s_c = s_c + 1$ of each cluster center
   – Get per-center learning rate $l = s_c^{-1}$
   – Update the cluster center (gradient-step) $\mu_{c_i} = (1 - l)\mu + lx_m$

It has been shown that the mini-batch algorithm [22, 23] speeds up the clustering process even with same or better accuracy than the standard k-means algorithm. The mini-batch algorithm requires the user to define the number of output clusters a-priori which makes the algorithm itself not applicable, especially in dynamic scenarios with over time growing data or the demand of having fully unsupervised clustering.

7.2.4 Chinese Whispers Algorithm (CW)

The Chinese Whispers algorithm [6] is a randomized graph-based algorithm, that is capable of clustering even larger datasets due to its linear complexity. In contrast to classical clustering algorithm it does not require any preliminary hyperparameter k and therefore can fully automatically be run providing similar or better results than classical vector-baased algorithms [6]. The simplicity of the CW algorithm itself together with the ability to be applied on weighted, unweighted, undirected or directed graphs provides a high range of use cases such as language separation or word sense disambiguation.

The general bottom-up approach of CW is straight forward as shown in the pseudocode below. Firstly, for each node a random class label is set. Then, iteratively the class labels are merged by examination of the local neighborhoods classes exhibiting the biggest sum of edge weights. If there are multiple identical winning classes a random class is chosen out of the set of possibilities. This process will continue unless the resulting clusters can be considered as stable.

CW algorithm

Input: The graph $G(V, E)$ where each node $v \in V$ and edge $e \in E$
Output: A set of clusters

a. **Initialisation** for all nodes $v_i \in V$ set a classlabel $c(v_i) = i$
b. **Loop**: while classlabel changes
c. **For all**: random $v \in V$ in
    i set class(v) = class with highest sum edge weights class in neighborhood of v

Apart from the many advantages of the CW algorithm it also comes with disadvantages: Due to its randomized properties, which makes the result change even for the same input data, it is not applicable in real world scenarios where the input data changes over time or the same output is required to quantify the result. Additionally the CW tends to form very small clusters, requiring the users intervention if only the dominant clusters are of interest. A re-clustering, i.e. by merging the unwanted small clusters into larger clusters, is not available and also due to the randomized properties of CW hardly to implement.

7.3 Novel Graph-Based Clustering Algorithms

7.3.1 Dynamic Clustering for Segregation of Co-Occurrence Graphs (DCSG)

DCSG is a novel graph-based clustering algorithm [12] following the learning process of the human brain. Similar to the ongoing learning process of a child becoming an adult it learns new terms and continuously categorization into related topics. Technically the process of learning is imitated by reading reading each document on sentence base and building a co-occurrence graph that represents the current state of knowledge. Each node therefore represents the learned term and the edges reflect the relation between different terms. For each novel term added to the co-occurrence graph, its distance to existing clusters represented by a centroid term [24] is measured using the inverse DICE [25] coefficient. The centroid term of each cluster is termermed as described in [24] by the node with the shortest average distance to every other node in the co-occurrence graph.

The clustering process involves two major steps, insertion and clustering, that behave differently depending on the case if the term to be added already exists within the graph or not. In the first case of a new term to be inserted, a new term represented by a node is created. In addition, the according edges are added to its co-occurrent terms. Then, the actual clustering step is performed, by merging the new term t_{new} into existing clusters, if the distance $t_{new} \leq \Delta d + 3\mu$. In case that $t_{new} > \Delta d + 3\mu$ a new cluster will be created. In addition, the algorithm will check if due to the insertion any relations may have changed in order to update the cluster centers correctly.

In case of an existing term to be inserted only the weight of the related egdes to other nodes will change. In addition, the cluster centers may change together with $\Delta d + 3\mu$. The membership of the according terms with that cluster therefore needs to be re-evaluated in case of exceeding the distance threshold might be moved to different clusters.

In the work of [12] it could be shown that the DCSG algorithm works especially on growing corpora very well without the need of a preliminary hyperparameter k, such as in the k-means algorithm. A general drawback on performance of DCSG is the initial learning phase in which the graph is not stabilized: in this case the cluster centers change frequently there is a high amount of recalculation steps required. As soon as the graph will stabilize, this will be compensated and the required recalculations will decrease.

7.3.2 *Sequential Clustering using Centroid Terms (SeqClu)*

The SeqClu algorithm [11] is a graph-based, hard-clustering algorithm that groups similar documents into clusters by using the concept of centroid terms [24, 26, 27].

It is assumed that the corpus to be clusters will grow over time. Therefore SeqClu processes each document sequentially one after another. Thus, each newly arriving document will be compared against an existing set of clusters containing previously clustered documents. The membership of each document to a certain cluster is evaluated by distance determination process that makes it descision based on a membership threshold. If a closest cluster can be found without exceeding a certain threshold the document will join the according cluster. If not, a new cluster will be formed by the document extending the cluster model.

Seqclu-Algorithm (antipodian version)

1. **Input:** A set of documents $D_i,\ i \geq 2$
2. **Output:** A set of document clusters $C_n,\ n \geq 2$
3. **Algorithm:**

    a. **For each** existing document D_i $(i \geq 2)$ determine its centroid term $\zeta(D_i)$ using spreading activation [28]

b. **Loop:** Determine the shortest path between the current document D_i and all of the other documents; recognize the most distant two documents D_a and D_b
c. **Initalisation:** Let D_a and D_b form the two initial clusters $C_{n=1} = \{D_a\}$ and $C_{n=2} = \{D_b\}$.
d. **Loop**: Determine for each additional or newly arriving document D_i its centroid $\zeta(D_i)$ in G and:
   i Calculate the membership of the current document D_i for each existing cluster $C_{n\geq1}$.
   ii Determine the winning cluster as the one for which D_i assumes the highest membership value.
   iii **If** the value of D_i's membership in the winning cluster does not exceed a given threshold,
   then associate D_i with the winning cluster $C_{win} \in C_n$
   else extend the model by a new cluster $C_{new} = \{D_i\}$

There are mainly three crucial parts that influence the behaviour of the algorithm:

- **Initialisation**
  The sequential clustering algorithm offers two main variants on choosing the initial clusters that influence the overall clustering quality as shown in [11]. In its basic variant the first arriving document—or a randomly chosen document if more than one initial document is available—is chosen. This may result, together with a poorly chosen membership threshold, into a large single cluster or overall inaccurate clustering results. Instead, as outlined in the pseudocode of SeqClu above, it is suggested to use the "antipodean" initialisation, that requires at least two documents available in order to determine the most distant documents forming the two initial clusters. Apart from the advantage of having better clustering results, the disadvantage of this approach is the probably high amount of calculations required to determine the two most distant documents. As with the work of Kubek [26] it is known that the co-occurrence graph will converge at about 100 documents processed. It is therefore suggested to consider only 2 to 100 documents during initialisation.
- **Cluster membership value**
  Each document must be assigned to an existing or new cluster. Thus, for each new document the average distance between the new documents centroid and all existing documents in each of the clusters is determined. As many path calculations may re-occur over time a caching mechanism is used to speed up the process of membership determination.
- **Winning cluster determination**
  Finally it must be decided wether a new document shall be merged into an existing cluster or a new cluster has to be created. This step is performed by using a dynamically for each cluster determined threshold value based on the local connections between the clusters centroid and its nearest neighbours. In order to reduce the amount of computation time, only neighbours within a certain radius are considered using a breadth-first search with limited depth.

Table 7.1 Conceptual differences of examined algorithms

Algorithm	Algorithm-type	Hyperparameter	Adaptive
K-means	Vector-based	k	No
K-means++	Vector-based	k	No
Minibatch	Vector-based	k	No
CW	Graph-based	None	No
SeqClu	Graph-based	None	Yes
DCSG	Graph-based	None	Yes

With the work of [11] it is shown that the SeqClu provides qualitative good clustering results on growing corpora without further need of a hyperparameter k. The main disadvantage of the SeqClu is the bad performance especially during the initialisation phase. Especially the variant of using antipodean documents is therefore subject of further research.

Conceptual differences of tested algorithms
From conceptual view the examined clustering algorithms show several differences that can greatly influence the usability and field of application of the clustering algorithms (Table 7.1). The graph-based clustering algorithms do not require any preliminary hyperparameter and therefore can be run completely autonomous even on dynamically changeling corpora. This property provides an great enhancement in uses-cases, such as web-engines, where the corpus over time may shrink or grow due to the availability of websites. Especially the inflexible requirement of a preliminary hyperparameter k prevents the use of classical vector-based algorithms in those cases. Furthermore a users manual intervention for every change in the data corpus seems unacceptable.

7.4 Experiments

7.4.1 Setup

Pre-processing tasks
Preliminary to all performed experiments classical natural language preprocessing such as

- sentence extraction,
- baseform reduction and
- stop word removal.

has been utilized on all documents. The resulting standardized data then were transformed into the algorithm dependent format. Related to the vector based algorithms, the TF-IDF-matrix [31] was created based on the documents term-vector available after natural language preprocessing. With reference to the graph-based clustering algorithms, the co-occurrence graph has been build where the nodes of the graph represent the terms and the weighted edges the significance of the co-occurring terms. The significance measure used is the DICE-coefficient [25], whose reciprocal value can be interpreted as the distance of the co-occurring terms. In addition, the co-occurence graph was adjusted by removing all small sub-graphs unless ony one large connected graph remained.

Technically the classical vector-based algorithms were implemented using Python with Scikit [33]. All other algorithms were implemented using Java and Neo4j [29].

Used corpora

The conducted experiments were performed multiple times over in size varying corpora of the German newspaper "Der Spiegel" (online version). For evaluation purpose each document was tagged according to its category provided by the author as an reference standard. The examined corpora itself contained 40 to 280 (size increasing by 20 documents) of randomly equally distributed selected documents of four categories. The chosen categories were *politics*, *cars*, *money* and *sports*.

Parametrization of tested algorithms

The classical vector based algorithms were parameterized as follows:

- K-Mean et al
  Four categories (hyperparameter k); 100 iterations; random cluster center
- Minibatch
  Four categories (hyperparameter k); 100 iterations; random cluster center; batch size: 100; initial samples: 300

The graph-based algorithms were parameterized as follows:

- Chinese Whispers
  100 iterations; no mutation
- SeqClu
  dynamic threshold; antipodean documents for initialisation
- DCSG
  No initialisation/hyperparameter required.

7.4.2 *Results and Discussion*

Clustering Quality

The evaluation of clustering algorithms can be considered as an complicated task due to the manifold inherent applications of clustering algorithms and use-case dependent interpretability [2, 30]. As the examined graph-based clustering algorithms provide a variable number of clusters, many external evaluation measures, e.g. f-measure or

Table 7.2 Average purity for each of the tested algorithms

Algorithm	K-Means	K-Means++	Minibatch	Chinese whispers	SeqClu	DCSG
Avg. total purity	0.79	0.83	0.76	0.77	0.72	0.71

Table 7.3 Comparison of purity for DSCG and SeqClu (different corpora sizes)

# of doc.	40	60	80	100	120	140	160	180	200	220	240	260	280
SeqClu	0.69	0.64	0.66	0.69	0.70	0.72	0.73	0.75	0.74	0.75	0.75	0.74	0.75
DCSG	0.65	0.67	0.67	0.65	0.71	0.70	0.72	0.75	0.71	0.73	0.77	0.71	0.72

rand index, are not applicable or falsify the evaluation result. The authors therefore decided to use the very simple purity measure [32], as it does not penalizes cases in which the number of output clusters is greater than the available classes.

Table 7.2 shows the accumulated purity for each algorithm examined based on corpora-sizes between 40 to 280 documents. It can be observed that the average purity of the classical algorithms resides between 0.7 and 0.83. Chinese Whispers also provides similar results having a purity of 0.77 in average.

The purity of the graph-based clustering algorithms SeqClu and DSCG exhibit a dependency on the number of documents which is shown in Table 7.3.

The dependency of purity on the amount of documents used can be explained by the general behaviour of the co-occurrence graph, which is known to converge at about 100 input documents. At this amount of documents the graph begins to converge and the clustering results, that depend on the centroids within the graph, will also become stable. A similar observation can be made for the DSCG algorithm that makes use of the inherent knowledge within the graph converging over time.

From Figure 7.1 it can be observed, that all clustering algorithms provide an overall similar good clustering quality at a purity-range from 0.7 to 0.8. The graph based algorithms exhibit the already mentioned behavior to stabilize at a purity around 0.71 at an amount of greater equal 100 documents.

Cluster-size and content

As in focus of the researcher only the SeqClu algorithm was examined more in detail regarding the cluster-sizes and plausibility of the clusters content. Regarding the cluster-size it could be observed that the average cluster-size is almost constant at 4 to 5 documents per cluster. Related to this observation is an slowly increasing number of clusters, which might indicate that there needs to be some adjustment mechanism depending on the overall corpus-size for SeqClu. As all experiments were performed using standard parametrisation this will be subject for future research.

As within the direct focus of the autors research, for SeqClu it could be observed that the clusters can be categorized in two main categories: homogenous and heterogeneous Clusters. The homogenous clusters contains only documents that are also

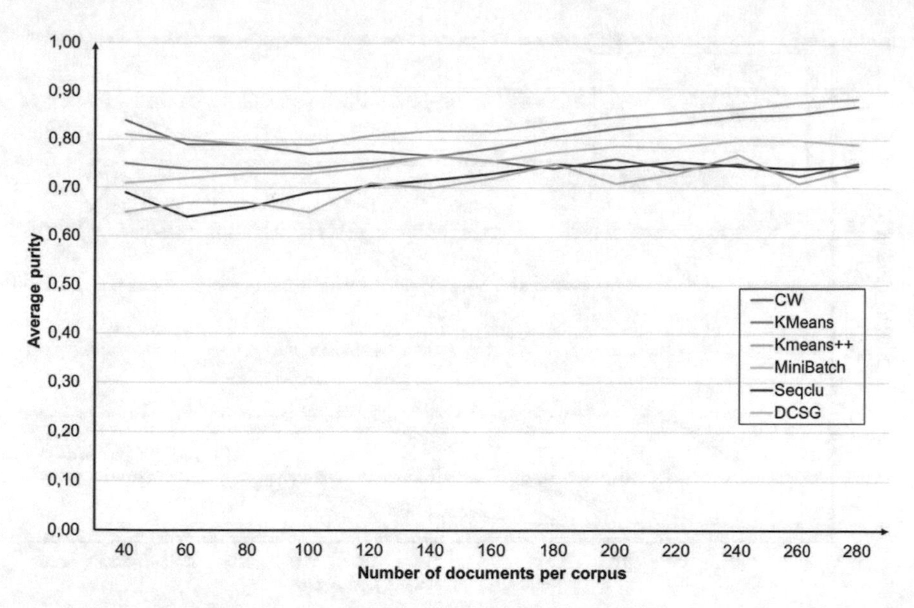

Fig. 7.1 Comparison of average purity for different corpora-sizes

matching the gold standards categorization. Most of these clusters were primarily self-contained and could be clearly distinguished from other topics. For example, documents related to the topic "Cars" could be clearly distinguished from the topic "Election" in Politics.

Heterogeneous clusters could be observed more frequently and in different graduations. Most of the observed heterogeneous clusters were actually almost homogenous having only a single document that could not be clearly assigned to an exact category. These document mainly consisted of features that were highly related to the clusters main topic but also exhibiting mayor aspects related to other topics. These clusters therefore might be considered as border clusters that mark areas within the co-occurrence graph changeling from one topic to another. In addition, for a few clusters it could be observed that they consisted of a mix of all available document categories, mainly containing documents can not be clearly assigned to any category. For some of the clusters it could be observed that from topical point of view the documents are often related to very generic topics such as "Money", which is also often part of other topics like Sports and Cars. Especially the heterogeneous clusters are subject of further research, e.g. on considering to change SeqClu into a softclustering algorithm or further optimization steps on the existing algorithm.

Clustering performance

For all algorithms the pure clustering execution time was compared against each other (Fig. 7.2) for different corpora sizes. From the shown results it can be observed that the classical clustering algorithms are in general faster than the graph-based clustering algorithms. Especially the Minibatch algorithm outperforms in comparison to other

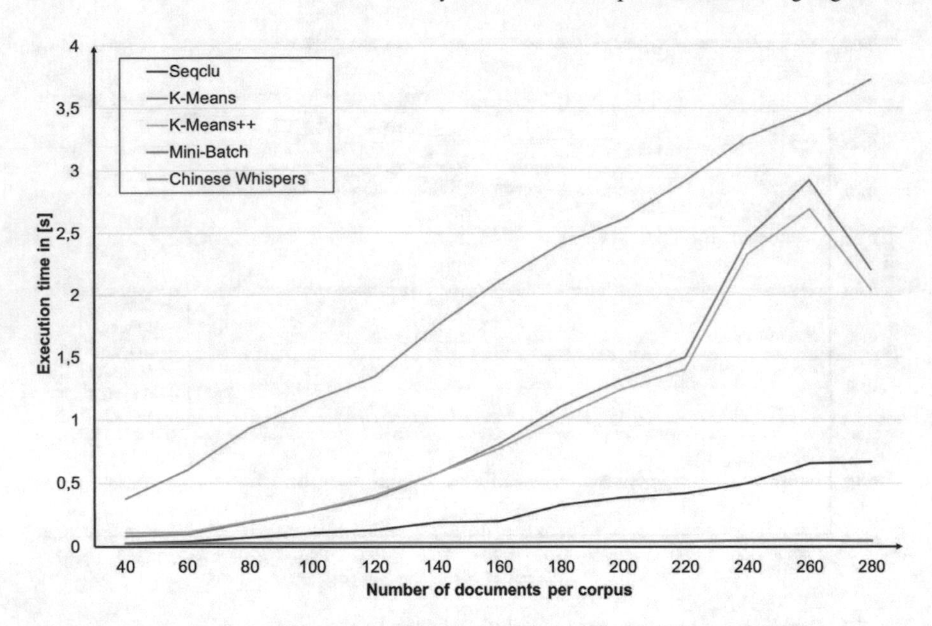

Fig. 7.2 Pure execution time for clustering

algorithms also on large corpora as it is designed to enhance the execution time in those cases.

Regarding the pure clustering time SeqClu exhibits a linear execution time and is also very fast in comparison to other algorithms. This is due to most recently implemented caching mechanisms on shortest path calculations during the initialisation phase. DCSG has been taken out of consideration for the pure execution time as its current implementation is very slow.

In addition to the measurement of pure clustering time for DCSG and Seqlu also the complete execution time (initialisation and clustering) was measured (Fig. 7.3).

In general it can be concluded that both algorithms performe too slow for real-time applications in cases where the initialisation has to be performed fast. For DCSG the main bottleneck to be identified is its sentence based approach which requires a huge amount of distance calculations. As the scope of DCSG is not to be tremendously fast but trying to find an accurate representation of knowledge this performance drawback is not essential for the entire quality of the algorithm.

Improvement of SeqCLu

The execution time of the sequential clustering algorithm depends on the calculation of the nearest neighbors and number of documents used during the determination of initial clusters. To improve the general performance of the algorithm, the most promising starting point is the initialisation phase whose calculation time is directly depending on the number of documents used.

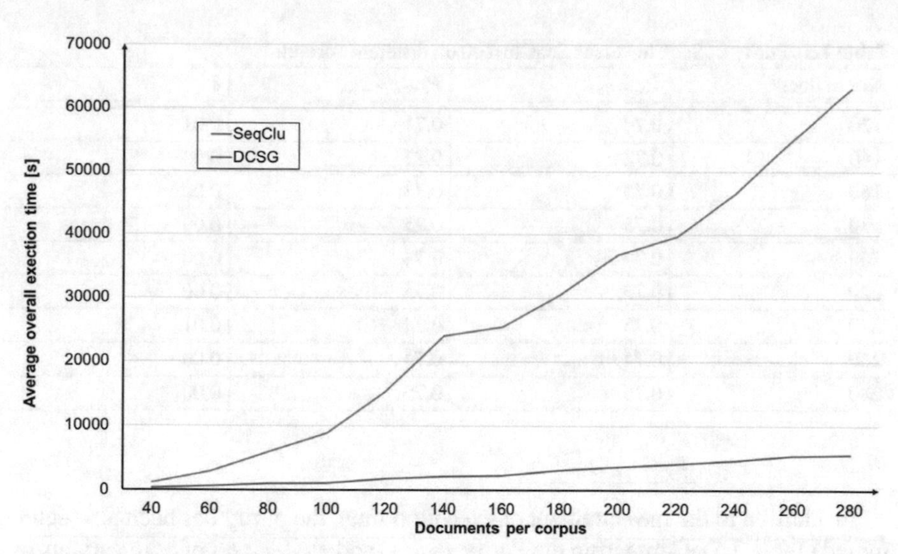

Fig. 7.3 Execution time of SeqClu and DCSG including initialisation and clustering

Table 7.4 Execution times SeqClu versus SeqClu-100 on different coropra

No. of doc.	$t_{e,seqclu}[min]$	$t_{e,seqclu-100}[min]$	$\delta_e[min]$
120	28.98	22.99	6.0
140	35.40	27.70	7.7
160	44.49	31.78	12.7
180	51.44	37.52	13.9
200	60.23	40.02	20.2
220	68.81	50.10	18.7
240	77.14	53.97	23.2
260	90.15	58.65	31.5
280	91.79	66.61	25.2

From preliminary experiments [26] it is known, that co-occurrence graph converges after processing approx. 100 documents. Thus, it can be concluded that beyond this amount of initial documents the co-occurrence graph can be considered as stable. In cases where more than 100 documents are initially available, only a random set of 100 documents should be considered for the initialisation process. This variant of SeqClu is referred as SeqClu-100.

In order to verify the above mentioned approch for all corpora the execution times for SeqClu versus SeqClu-100 have been determined and compared. From Table 7.4, which compares the execution time of SeqClu against SeqClu-100 and saved time δ_e (difference in execution time), it can be observed that the overall execution time could be greatly improved.

Table 7.5 Purity of SeqClu versus SeqClu-100 on different coropra

No. of doc.	P_{SecClu}	$P_{SecClu-100}$	δ
120	0.70	0.71	0.01
140	0.72	0.72	0.00
160	0.73	0.73	0.00
180	0.75	0.75	0.00
200	0.74	0.74	0.00
220	0.75	0.75	0.00
240	0.75	0.74	0.01
260	0.74	0.74	0.00
280	0.75	0.75	0.00

In addition to the measurement of execution time, the purity has been also determined (Table 7.5). Comparing the Purity P_{SecClu} and $P_{SecClu-100}$ of both variants by determination of $\delta = |P_{SecClu} - P_{SecClu-100}|$ it can be easily observed that there is no negative impact using 100 randomly chosen documents during the initialisation of SeqClu.

Additional experiments also showed that the average cluster-sizes and number of output clusters do not change significantly.

7.5 Conclusion

Within the provided work a comparison between vector- and graph-based algorithms was made. Both categories of algorithms were compared regarding their performance and clustering quality in the field of document clustering. It can be concluded that the vector-based algorithms in general perform at a higher speed in contrast to graph-based clustering, that typically depend on more expensive operations over the co-occurrence graph.

The requirement of having a-priori definition of the hyperparameter k forces the user's intervention and limits the algorithms application to scenarios where the corpus is not changing over time. The graph-based clustering algorithms in contrast show that a good categorization without the preliminary requirement of a hyperparameter is possible. One biggest advantage over classical clustering is the property of having an associative representation—similar to the human brain—that provides a better differentiation of topical areas within the graph and according clusters. By investigating SeqClu's clustering results in depth new knowledge was gained in order to optimize the clustering quality by investigating more closely into the clusters structure and relations to each other. In addition, it could be shown that SeqClu can be significantly improved performance wise by limiting the number of documents during initialisation without having negative effects on the overall clustering quality.

References

1. Estivill-Castro, V.: Why so many clustering algorithms: A position paper. SIGKDD Explor. Newsl. **4**(1), 65–75 (2002)
2. Wierzchon, S., Kłopotek, M.: Modern Algorithms Cluster Anal. **34**, 01 (2018)
3. MacQueen, J.: Some methods for classification and analysis of multivariate observations. In: Proceedings of the Fifth Berkeley Symposium on Mathematical Statistics and Probability, Volume 1: Statistics, pp. 281–297. Berkeley, Calif. (1967). University of California Press
4. Arthur, D., Vassilvitskii, S.: K-means++: The advantages of careful seeding. In: Proceedings of the Eighteenth Annual ACM-SIAM Symposium on Discrete Algorithms, SODA '07, pp. 1027–1035. Philadelphia, PA, USA (2007). Society for Industrial and Applied Mathematics
5. Altman, N.S.: An introduction to kernel and nearest-neighbor nonparametric regression. Am. Stat. **46**(3), 175–185 (1992)
6. Biemann, C.: Chinese whispers: An efficient graph clustering algorithm and its application to natural language processing problems. In: Proceedings of the First Workshop on Graph Based Methods for Natural Language Processing, TextGraphs-1, pp. 73–80. Stroudsburg, PA, USA (2006). Association for Computational Linguistics
7. Kubek, M., Unger, H.: Towards a librarian of the web. In: Proceedings of the 2Nd International Conference on Communication and Information Processing, ICCIP '16, pp. 70–78. ACM, New York, NY, USA(2016)
8. Christen, M. et al.: YaCy: Dezentrale Websuche (2017). Online Documentation on http://yacy.de/de/Philosophie.html
9. Kubek, M.: Dezentrale, kontextbasierte Steuerung der Suche im Internet. PhD thesis, Hagen (2012)
10. Kubek, M.: Concepts and Methods for a Libarian of the Web. FernUniversität in Hagen (2018)
11. Hloch, M., Kubek, M.: Sequential clustering using centroid terms. In: Autonomous Systems 2019: An Almanac, pp. 72–88. VDI (2019)
12. Simcharoen, S., Unger, H.: Dynamic clustering for segregation of co-occurrence graphs. In: Autonomous Systems 2019: An Almanac, pp. 53–71. VDI (2019)
13. Bacila, M., Adrian R., Ioan, M.: Prepaid telecom customer segmentation using the k-mean algorithm. Analele Universitatii din Oradea **XXI**, 1112–1118 (2012)
14. Zulfadhilah, M., Prayudi, Y., Riadi, I.: Cyber profiling using log analysis and k-means clustering a case study higher education in Indonesia
15. Aggarwal, C., Reddy, C.: Data Clustering Algorithms and Applications. Chapman and Hall (2013)
16. Pena, J.M., Lozano, J.A., Larranaga, P.: An Empirical Comparison of Four Initialization Methods for the k-Means Algorithm (1999)
17. Forgy, E.: Cluster analysis of multivariate data: Efficiency versus interpretability of classification. Biometrics **21**(3), 768–769 (1965)
18. Hartigan, J.A., Wong, M.A.: Algorithm AS 136: A K-Means clustering algorithm. Appl. Stat. **28**(1), 100–108 (1979)
19. Khan, S., Ahmad, A.: Cluster center initialization algorithm for k-means clustering. Pattern Recognit. Lett. **25**, 1293–1302 (2004); Int. J. Adv. Comput. Sci. Appl. **7**, 08 (2016)
20. Kaufman, L., Rousseeuw, P.: Finding Groups in Data: An Introduction to Cluster Analysis (1990)
21. Caliński, T., Harabasz, J.: A dendrite method for cluster analysis. Commun. Stat. Theory Methods **3**, 1–27 (1974)
22. Sculley, D.: Web-scale k-means clustering. In: Proceedings of the 19th International Conference on World Wide Web, WWW '10, pp. 1177–1178. New York, NY, USA (2010). Association for Computing Machinery
23. Feizollah, A., Anuar, N., Salleh, R., Amalina, F.: Comparative study of k-means and mini batch k- means clustering algorithms in android malware detection using network traffic analysis (2014)

24. Kubek, M., Unger, H.: Centroid terms as text representatives. In: Proceedings of the 2016 ACM Symposium on Document Engineering, DocEng '16, pp. 99–102. ACM, New York, NY, USA (2016)
25. Dice, L.R.: Measures of the amount of ecologic association between species. Ecology **26**(3), 297–302 (1945)
26. Kubek, M., Böhme, T., Unger, H.: Empiric experiments with text representing centroids. In: 6th International Conference on Software and Information Engineering (ICSIE 2017) (2017)
27. Kubek, M., Unger, H.: Centroid terms and their use in natural language processing. In: Autonomous Systems 2016. VDI-Verlag Düsseldorf (2016)
28. Kubek, B., Unger, H.: Spreading Activation: A Fast Calculation Method for Text Centroids (2017)
29. Vukotic, A., Watt, N., Abedrabbo, T., Fox, D., Partner, J.: Neo4j in Action. Manning (2015)
30. Everitt, B.S., Landau, S., Leese, M.: Cluster Analysis, 4th edn. Wiley Publishing (2009)
31. Salton, G., Wong, A., Yang, C.S.: A vector space model for automatic indexing. Commun. ACM **18**(11), 613–620 (1975)
32. Christopher, D.: Manning, Prabhakar Raghavan, and Hinrich Schütze. Introduction to Information Retrieval. Cambridge University Press, USA (2008)
33. Scikit-learn: Machine Learning in Python. https://scikit-learn.org/
34. Biemann, C., Quasthoff, U., Heyer, G., Holz, F.: ASV toolbox: a modular collection of language exploration tools. In: Proceedings of the Sixth International Conference on Language Resources and Evaluation (LREC'08). Marrakech, Morocco (2008). European Language Resources Association (ELRA)

Chapter 8
A Neighbourhood-Based Clustering Method for Graph Data Models

Santipong Thaiprayoon and Herwig Unger

8.1 Introduction

Due to the rapid evolution of digital information on the Internet, such as text, audio, images, and video, a mechanism for automatically clustering and extracting those data objects is necessary. Clustering has attracted great attention both in academia and industry due to its potential applications in data and text mining [1]. The mechanism of clustering is an unsupervised machine problem that involves grouping the population or data points together into a set of classes called clusters (groups or categories) in the data space. Each cluster contains the most similar data points and is dissimilar in distinctive clusters. The basic assumption is that the data points within a cluster differ in the same cluster from the data points of other clusters [2]. In contrast, clustering has been widely applied in data analysis by a graph, which is one of the most popular research areas.

Graph clustering algorithm refers to the clustering of data objects in the form of graphs aiming to group a set of similar nodes in a graph into disjoint clusters. Moreover, clustering algorithms have become a useful technique for unsupervised feature selection and dimension reduction, especially when dealing with sparse data problems [3]. There are several tasks in Natural Language Processing (NLP) to apply graph-based models to handle textual data. One of the important applications is word clustering. It is one of the most fundamental text mining tasks for finding similar hierarchies of words or concepts [4].

Word clustering is one of the most fundamental text mining tasks that aims at dividing a set of words into meaningful clusters. This technique is an essential task for reducing a large number of text documents and obtaining meaningful information, which leads to significant improvements in real-world applications in various domains of natural language processing such as information retrieval, document clustering, and recommendation system. For example, in information retrieval, a word clustering algorithm could help discover related words as query expansion around search queries, which enhances the precision of a query and displays search results accurately. Moreover, it is the practical benefit for clustering the web search results

H. Unger and M. Kubek, *The Autonomous Web*, Studies in Big Data 101,
https://doi.org/10.1007/978-3-030-90936-9_8

returned by a search engine. In document clustering, a word clustering algorithm is a fundamental step of the document clustering process. It is important to build text representations from documents that are vulnerable to measure the similarity between documents.

However, word clustering is still a challenging task in finding the optimal number of clusters and dividing similar words into meaningful groups. The basic idea of the proposed mechanism is imitated from the learning process in the brain of humans inspired by neuroscience [5] on modular segregation for determining the clusters of knowledge. The brain starts learning new words from text documents and builds categories for the knowledge. A word co-occurrence graph is employed to represent the brain of humans. The set of nodes is created from the set of words. The edges are established by pairs of words read in the same sequence in text documents.

To overcome this problem, this paper proposes a new mechanism of clustering of words in a co-occurrence graph based on a graph model. The proposed mechanism aims to discover similar words and assign them to the same cluster. The isolated subclusters are then iteratively combined according to a value of the distance range until convergence criteria or no words are moved to other clusters, which aim to reduce the number of clusters. To calculate the distance between words in a graph, Dijkstra's algorithm is utilized to measure the shortest paths between words in a graph. The distance from a source word to a destination word is calculated as the cost of traversing between two words in a co-occurrence graph. The total of the lowest cost is used for identifying the cluster centroid in a cluster [6]. Finally, the result of the proposed mechanism is a collection of word clusters.

The main contribution of this paper consists of two parts. Firstly, an unsupervised word clustering using a co-occurrence graph is proposed. The method automatically groups a set of word clusters without any initialization and parameter selection. Moreover, this mechanism does not require a certain number of clusters. Secondly, this mechanism also plays an essential role in several tasks of natural language processing. The key idea is that a collection of word clusters is applied to deal with dimensionality reduction, computational complexity, and improve efficiency and accuracy on real-world applications.

The remainder of this paper is organized as follows. In the next section, fundamental theories are reviewed, and the proposed method is discussed in great detail. The experimental results and discussions are illustrated in Sect. 8.3. Finally, the conclusions and suggestions for future works are presented.

8.2 Clustering Mechanism

In this section, several major topics are described as fundamental concepts for developing the proposed mechanism, including co-occurrence graph construction, and the basic definitions of finding centroid terms. Next, the mechanism of cluster building on a co-occurrence graph is explained.

8.2.1 Conceptual Approach

The key objective of the proposed mechanism is to automatically build a set of clusters of words based on their word similarity with a conceptual representation of the co-occurrence graph. The result of the proposed mechanism is a collection of word clusters that contains the most similar words which share a similar concept or semantic meanings. The process overview of the proposed mechanism is that the text documents are converted into a set of word co-occurrences within a given sliding window of size. The text processing technique, including stop words removal, word normalization, and nouns extraction is performed to select only nouns and filter out noisy information or unimportant words in the documents. The set of word co-occurrences is then inserted into the co-occurrence graph where the words are the nodes and the co-occurrences are the connections between them. The mechanism of cluster building is applied to assign each word to a proper cluster according to a distance range. Then, recalculating the cluster centroid of newly formed clusters. To merge similar clusters becoming the same cluster, subclusters on the co-occurrence graph find the most similar clusters and merge them into a new bigger cluster based on the distance range between the centroid of clusters. Finally, recalculating the centroid of newly formed clusters again, which also helps to reduce the number of clusters and fit them into another cluster.

8.2.2 Co-Occurrence Graph Construction

The construction of the co-occurrence graph is a basic model to store information of text documents as graph-based text representation [7]. Graph-based text representation is capable of representing the semantic and structural information of the text document effectively. The idea is that words that appear in the same context together in a text document will have similar meanings. A text document is performed using the text processing method to select important words and convert them into a set of candidate words. A co-occurrence graph is constructed to define similarity among words in a sentence within a fixed-size sliding window, where V is a set of nodes represented as unique words. E is a set of edges representing the relationship between a pair of words. The weight of the relation between two words is added to the corresponding edge in the co-occurrence graph G. The edge is calculated by the raw frequency of occurrence. The weight on the edge shows the significance of the association between words in the co-occurrence graph. The well-known measurements of importance calculations are the dice coefficient, mutual information, frequency of occurrence, and the log-likelihood ratio.

8.2.3 Determination of Cluster Centroids

A centroid is a key term that can be used to represent text as the center of a document. To identify centroid terms of the text document, a co-occurrence graph is used [8]. The distance $d(w_a, w_b)$ between two words w_a and w_b in the graph G can be defined by as the reciprocal of the significance value:

$$d(w_a, w_b) = \frac{1}{sig(w_a, w_b)} \tag{8.1}$$

Consequently, the distance of two words being isolated words or situated in two, not connected sub-graphs is set to

$$d(w_a, w_b) = \infty \tag{8.2}$$

A centroid term of a document is the term with the minimum average distance to all words of the document in the co-occurrence graph.

$$d(w_a, w_b) = \frac{\sum_{i=1}^{N} d(w_i, t)}{N} = MIN \tag{8.3}$$

The centroid terms can be used to determine the semantic distance of documents. The method, therefore, can be applied as a text representation of the text document in tasks of natural language processing such as text clustering, text categorization, and text classification.

8.2.4 Cluster Building

This component is based on a fundamental concept of assigning each word to a cluster, which was presented in previous work consisting of two main modules. Figure 8.1 illustrates the overall process of the construction of the co-occurrence graph and word clusters, which can be explained in the details below.

The basic idea of this method is derived from the learning process in the brain of humans. The brain begins to learn new words and build categories for this knowledge. This method applies a co-occurrence graph for clustering. The process of cluster building is that one document is read sentence by sentence. Each word of each sentence and relation (distance) between words are added. Then, a co-occurrence graph is created. The clustering process is performed to assigned words into a cluster depending on the distance range to the cluster center (centroid). The significance values are determined by using a word similarity metric. The cluster center is a word that has the minimum average distance to all words of the documents in the graph. A cluster center is generated when a new cluster is formed, it is updated when the cluster members increased, and it is used to calculate an average distance (μ) of all

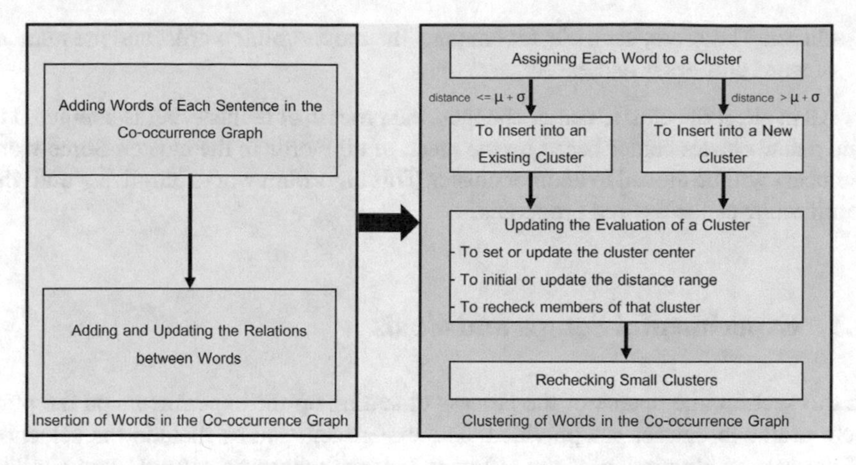

Fig. 8.1 The Process of the Construction of the Co-occurrence Graph and Word Clusters

words from the cluster center. Then, a standard deviation (σ) of the distance range is obtained in each cluster. The distance range of the cluster center is used to validate that the word is in the distance range ($\mu + \sigma$). In each insertion of a new word or change of an edge weight, the evaluation of the clustering change. Clusters can be joined, divided, and restructured. After the listing of documents is added sequentially, each sentence of each document is processed as follows:

1. Insertion of words in the co-occurrence graph: Each sentence is preprocessed, including stemming, stop words removal, and nouns extraction. A set of word pairs is then inserted into the co-occurrence graph. A relation between words is weighted by a distance score that represents the strength of the relationship of words.
2. Clustering of words in the co-occurrence graph: a word of each sentence must be assigned with a cluster. If a word is the first word of any cluster, then set this word to a cluster center and set the distance range. If the distance range is less than or equal to a pre-defined threshold, then insert the word into an existing cluster. If the distance is more than to pre-defined threshold then a cluster is formed as a new cluster. After assigning words into a suitable cluster, a formed new cluster is computed to determine a cluster center based on the minimum average distance of all words in the cluster, and then repeat all steps until the maximum number of iterations is reached.
3. Merging subclusters: the subclusters are small clusters or isolated words because some words may move to another cluster or create a new cluster. These clusters are performed to reduce the number of clusters by merging similar clusters to become the same cluster. Each small cluster is assigned to the closest cluster based on the minimum distance between words of cluster centroids according to the threshold of the distance range. If the distance range between two cluster centroids is greater than a pre-defined threshold. The cluster is not merged and then creating a new

cluster. Therefore, each cluster contains the most similar words, and the number of word clusters is decreased.

All in all, if the cluster center changes, the process of reclustering is computed to find a new cluster center based on the mean of all words in the cluster. Some word members will be moved to another cluster. This algorithm works iteratively until the result achieves the desired properties.

8.3 Experimental Setups and Goals

In this section, the details of the process of setting up the experiments on the publicly available dataset are provided. The experiment setups include the selection of datasets, configuration of the different distance ranges on several word similarity metrics, and calculation of cluster quality. The main objective is to measure the strength of the semantic relationship between words in a co-occurrence graph against other metrics, including the Dice coefficient, Positive Pointwise Mutual Information (PPMI), and co-occurrence frequency. The experimental setups and goals of the proposed mechanism are shown in subsections.

8.3.1 Goals

The proposed mechanism is evaluated in the experiments for building a set of word clusters on the co-occurrence graph. The objectives of this study are demonstrated as follows:

- To comparing against word similarity metrics for identifying the strength of the semantic relationship between words in a co-occurrence graph
- To determine the appropriate standard deviation of distance in different values from 1 to 3 for choosing the best value of the distance range.
- To evaluate the accuracy of the word clustering algorithm for identifying the cluster quality
- To explore the effectiveness and efficiency of the proposed mechanism

To prove the goals, the experiments are described in the next subsection.

8.3.2 Data Description

To evaluate the effectiveness and efficiency of the proposed mechanism, the British Broadcasting Corporation (BBC) news dataset was used for the experiments. The BBC news dataset was randomly selected by humans containing 100 news articles with technology domain from the year 2004–2005. This dataset was also prepared for evaluating the performance of the proposed mechanism as a ground truth dataset. The growth truth dataset was manually assigned by human judgments due to no public dataset suitable for word clustering evaluation.

8.3.3 Evaluation Metric

The purity metric was used to assess the precision of the proposed mechanism. The purity is an external evaluation metric that necessitates the use of a ground truth dataset. For the purity calculation, each cluster is assigned with the most frequent class label. The number of correct class labels in each cluster is the sum and divided by the total number of words. The purity that closes to 0 means a weak clustering, and a good clustering has a purity score of 1. The calculation of purity is defined by Eq. 8.4 as follows.

$$Purity = \frac{1}{N}\sum_{i=1}^{k} max_j |c_i \cap t_j| \tag{8.4}$$

Where N is the number of all words, k is the number of clusters, c_i is a cluster in a set of clusters, and t_j is the classification, which has the max count for a cluster c_i.

8.3.4 Experimental Setups

To show that the goals of the proposed mechanism are achieved, the experiments were conducted on the BBC news dataset focusing on finding an appropriate word similarity metric, and distance ranges according to the quality of the clustering. The empirical experiments were divided into two phases for evaluating the performance of the proposed mechanism. For the first phase of the experiment, the experiment was varied by the standard deviation on the same dataset. The standard deviation of distance was created in different values from 1 to 3 to determine the appropriate value that was used for determining membership in a cluster. A word pair was created by co-occurrence between two words within the size of the sliding window at 3 words. The ground truth dataset was considered to validate the effect on the purity measure. To identify the quality of clusters, the purity measure was employed like cluster cohesion and the number of clusters to evaluate the accuracy of the word clustering algorithm. For the second phase of the experiment, this phase was to compare against three similarity metrics for identifying the strength of the semantic relationship between words in a co-occurrence graph, including Dice coefficient, Positive Pointwise Mutual Information (PPMI), and raw frequency of occurrence of two words. All of the clustering mechanisms were performed on a personal computer (PC) with an Intel (R) Core (TM) i5-4570 CPU @ 3.20 GHz and 16 GB of RAM.

8.3.5 *Experimental Results and Discussions*

To ensure that the proposed mechanism in the co-occurrence graph based on graph-based models is suitable to be applied, the evaluation results are mainly described in the number of words and clusters with varying values of the standard deviation on different similarity metrics. This study examines an effect on different values of the standard deviation making the purity score possible to change in the co-occurrence graph. In addition to accuracy, the evaluation is reported on the purity measure. The experiment results of the clustering analysis are given below.

8.3.5.1 The Number of Words on the Dataset

This experiment aims to consider the number of words with the dataset in the co-occurrence graph. The experiment results are demonstrated in Fig. 8.2.

From Fig. 8.2, the graph shows the number of words was stable when the number of documents is nearly 100 documents.

8.3.5.2 The Number of Clusters on the Dataset

To ensure that the proposed mechanism could build the number of clusters efficiently, the paper conducted the experiments by varying the distance range to generate word clusters with the same corpus. The standard deviation of distance was created in different values from 1 to 3. The experiment results are presented in Fig. 8.3.

From Fig. 8.3, the graph shows the number of clusters on the difference of the distance range. The number of clusters increased significantly in every value of the distance range when the number of documents dramatically grew. The distance range $(\mu + w/o\ S.D.)$ had the highest number of clusters, while the other values of the distance range had fewer numbers respectively. The number of clusters on distance range $(\mu + w/o\ S.D.)$ differed from the others because the threshold of the distance

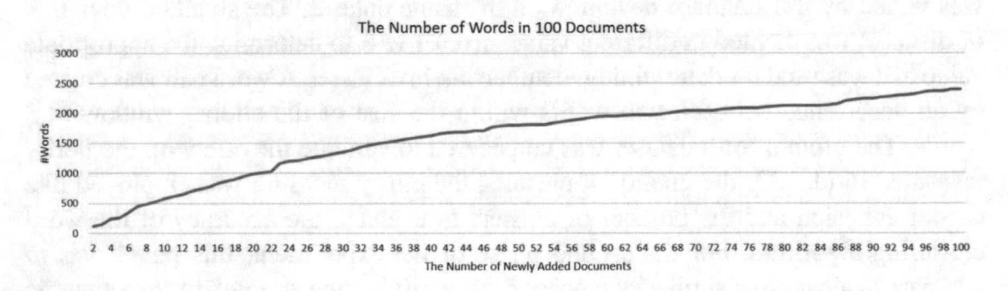

Fig. 8.2 The number of words in 100 documents

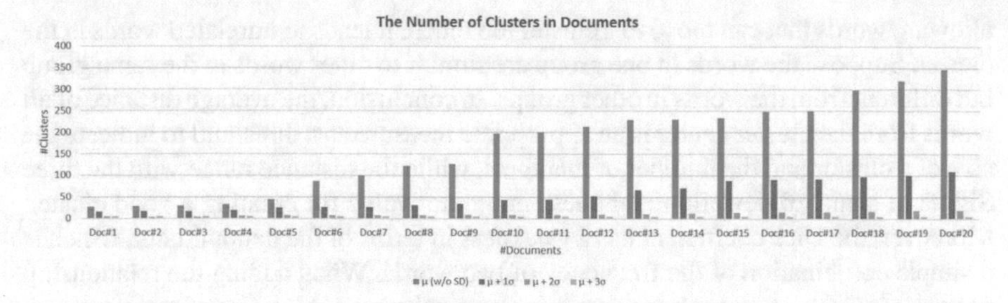

Fig. 8.3 The number of clusters on the dataset

range from the cluster centroid was less than the others, which resulted in the words that were could not be assigned to the clusters. When a word could not belong to the clusters, the word created a new cluster, leading to the number of clusters jumps substantially.

8.3.5.3 The Experimental Results with Varying Values of Distance Ranges on Different Similarity Metrics

To evaluate the accuracy and find the best distance range comparing with other similarity measures, the purity was calculated for assessing the quality of clusters. The comparison of different distance ranges on the different similarity metrics was examined in this experiment. The experimental results with different values of the standard deviation from 1 to 3 are summarized in Fig. 8.4.

From Fig. 8.4, the results indicated that the metric of the raw frequency of occurrence achieved high purity on the ground truth dataset. The distance without the distance range yielded better results in both the size of the cluster and the purity score. The reason is that adding the value of the standard deviation sounds like a good idea. The main objective of the addition of the value of the standard deviation is to expand the size of a cluster and allow words to move to the cluster more than ever. But in practical works, the standard deviation is unnecessary because when

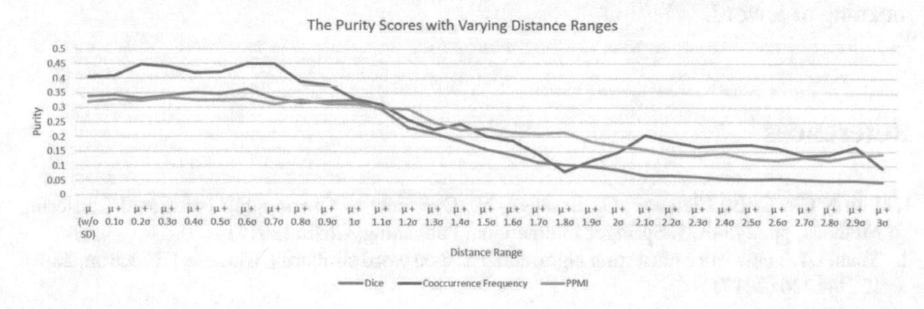

Fig. 8.4 The purity score with varying distance ranges

allowing words that can move to a cluster too much, it leads to unrelated words in the cluster. Suppose the words in one group are similar to other words in the same group but different from the words in other groups. In conclusion, the average distance of all words from the cluster center is an appropriate measure as a threshold to indicate the size of a cluster and the number of members, while the distance range with the three Sigma of standard deviation is the less important value for creating a good cluster. Moreover, the Dice coefficient has a weakness in terms of the denominator, which is a simple combination of the frequency of two words. When adding the relationship of words into the co-occurrence graph, it also results in the high frequency of a word, and it has a negative effect on the value of the Dice coefficient in the co-occurrence graph. Therefore, the denominator makes the value of the Dice coefficient possible to decrease because the numerator is less than the denominator. This means that the strength of the relationship between two words is also reduced.

The proofs and experimental results indicated that the proposed mechanism could group semantically similar words into the same cluster on the co-occurrence graph effectively and efficiently. In addition, the proposed mechanism is unsupervised learning and generic, which could be applied to various tasks in natural language processing and text mining. This is an important tool to address several real-world problems.

8.4 Conclusion

This paper introduces a new mechanism of automatic word clustering based on graph models, which aim to group related words into a proper cluster. The final result of the proposed mechanism is a collection of word clusters. The experimental results indicated that the proposed mechanism could efficiently divide similar words in the co-occurrence graph into the same cluster. This mechanism is thus the main tool that could help to solve several NLP tasks. For future works, the paper plans to apply word embeddings to calculate the semantic distance of words for improving the accuracy of word clustering algorithms in terms of contextual semantic meanings. Moreover, a named entity recognition is used to recognize named entities such as person names, locations, and organizations, which help reduce lexical ambiguity to understand the meaning of a word.

References

1. Ben N'Cir, C.-E., Cleuziou, G., Essoussi, N.: Overview of Overlapping Partitional Clustering Methods, pp. 245–275. Springer International Publishing, Cham (2015)
2. Yuan, L.: A new word clustering algorithm based on word similarity. Chinese J. Electron. **26**(6), 1221–1226 (2017)

3. Wenliang, C., Xingzhi, C., Huizhen, W., Jingbo, Z., Tianshun, Y.: Automatic word clustering for text categorization using global information. In: Information Retrieval Technology, pp. 1–11. Berlin, Heidelberg (2005). Springer Berlin Heidelberg
4. Biemann, C.: Chinese whispers—an efficient graph clustering algorithm and its application to natural language processing problems. In: Proceedings of TextGraphs: the First Workshop on Graph Based Methods for Natural Language Processing, pp. 73–80. New York City (2006). Association for Computational Linguistics
5. Baum, G.L. et al.: Modular segregation of structural brain networks supports the development of executive function in youth. Curr. Biol. **3**, 1561–1572 (2017)
6. Thaiprayoon, S., Unger, H., Kubek, M.: Graph and centroid-based word clustering. In: Proceedings of the 4th International Conference on Natural Language Processing and Information Retrieval, NLPIR 2020, pp. 163–168. New York, NY, USA (2020). Association for Computing Machinery
7. Kubek, M., Unger, H.: Centroid terms as text representatives. In: Proceedings of the 2016 ACM Symposium on Document Engineering, DocEng '16, pp. 99–102. New York, NY, USA (2016). Association for Computing Machinery
8. Unger, H., Kubek, M.: On evolving text centroids. In: Recent Advances in Information and Communication Technology 2018, pp. 75–82, Cham (2019). Springer International Publishing

Chapter 9
Decentralised Routing in P2P-Systems with Incomplete Knowledge

Supaporn Simcharoen, Gisela Nagy, and Herwig Unger

9.1 Introduction

The routing process is the task of selecting suitable paths to transfer information between nodes in the graph. Thus, the best path links from a source node to a destination node is always chosen. However, determining the lowest cost routing of a given source-destination pair requires a necessary consideration [1]. Moreover, selecting the shortest path of two given nodes in graph theory is a classic problem [2, 3]. Although Dijkstra's algorithm [4] can be applied to solve this problem, whereas, in route preparation of massive datasets, this algorithm processes extremely slow [5].

A co-occurrence graph, an influential underlying information structure, is constructed from collections of documents that represent knowledge [6]. A co-occurrence is the common occurrence of two words in a sentence. The analysis of co-occurrences makes it possible to find words with similar meanings. As these words are mentioned together more frequently, they are immediately found to be related to one another. While several documents are inserted, the potential of a single computer (called peer) and long processing times to manage words or nodes in co-occurrence graphs become problematic. A decentralised approach of creating co-occurrence graphs from text documents in different peers under the project name the web search engine 'TheBrain', is a potential solution to these problems. Besides, the system can achieve a high degree of parallelism and scalability.

TheBrain, the second version of WebEngine [7], is a purely distributed P2P System that works on several peers without a controlling authority and can be distributed across networks. In TheBrain, a large distributed co-occurrence graph and smaller local co-occurrence graphs as a limited user knowledge are constructed. These co-occurrence graphs are built from a set of word co-occurrences of the contents (documents). The local co-occurrence graphs will grow from these contents and will be more topically specialised over time (see Fig. 9.1).

The decentralised routing process is the task of selecting the most suitable path in the graph. In a peer, this process calculates a path based on its own existing

H. Unger and M. Kubek, *The Autonomous Web*, Studies in Big Data 101,
https://doi.org/10.1007/978-3-030-90936-9_9

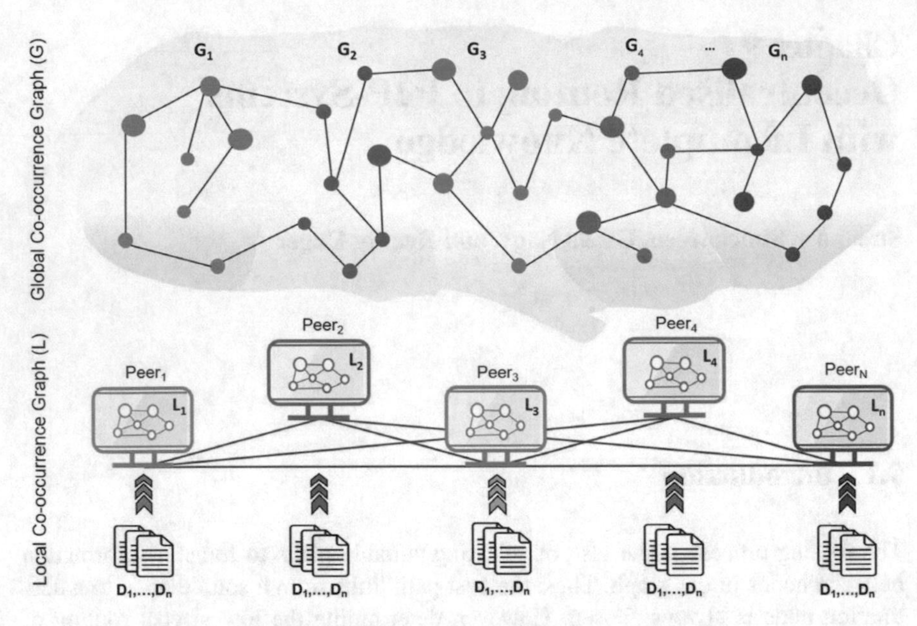

Fig. 9.1 Construction of co-occurrence graphs in TheBrain

knowledge and the knowledge of its neighbours. This path is calculated without the full view of the graph. An iterative process of calculation and transfer of information with neighbours are considered in this suitable route determination. This process can be compared to car drivers who generally do not know and remember all the routes. Their frequent routes will be only remembered as their limited knowledge. When they travel to a new destination, i.e., a place or a building where they have never visited before, asking someone for directions might be the solution. However, the travel information may be changed during the journey because travelling to that destination has many routes. Moreover, better and more suitable routes can be acquired from other informants along the way.

As a result, every peer has incomplete knowledge, and no words on the co-occurrence graph have complete routing information. For this purpose, a new routing mechanism is presented that allows decentralised routing with incomplete information. The routing analysis on the distributed co-occurrence graph will be considered to ensure that this routing mechanism is suitable to be applied to the search process. Moreover, the numbers of routing in the local and global co-occurrence graphs are compared. Furthermore, in order to determine that the decentralised routing can reduce the distances, all source-destination word pairs on the co-occurrence graphs are examined. In addition, this mechanism can be applied to optimise the distributed co-occurrence graph and reduce communication overhead.

9.2 Conceptual Approach

9.2.1 *Routing Mechanism with Incomplete Knowledge*

In each peer, a list of documents is assigned by a user i; the local and global co-occurrence graphs are built from a set of word co-occurrences of those documents. In a distributed co-occurrence graph $G = (W, E)$, two graphs are constructed (see Fig. 9.1):

- The global co-occurrence graph G_i: a small part of the global co-occurrence graph G.
- The local co-occurrence graph L_i: a local user co-occurrence graph as a limited knowledge.

The user i knows only words in the subgraph G_i and in the local co-occurrence graph L_i. The key idea of the local routing method is the user can assign a query q to which a path p from any known word w_i with $p = w_j, w_2, ..q$. It is certain that this path p also exists in the global co-occurrence graph G. The path and the neighbourhood information of G are sufficient to search for a routing path (see Fig. 9.2).

In the global co-occurrence graph G, the routing with the local co-occurrence graph is used in the word locating and updating processes. The routing analysis on the distributed co-occurrence graph is described in the next section. This is to ensure that this routing mechanism is suitable to be applied.

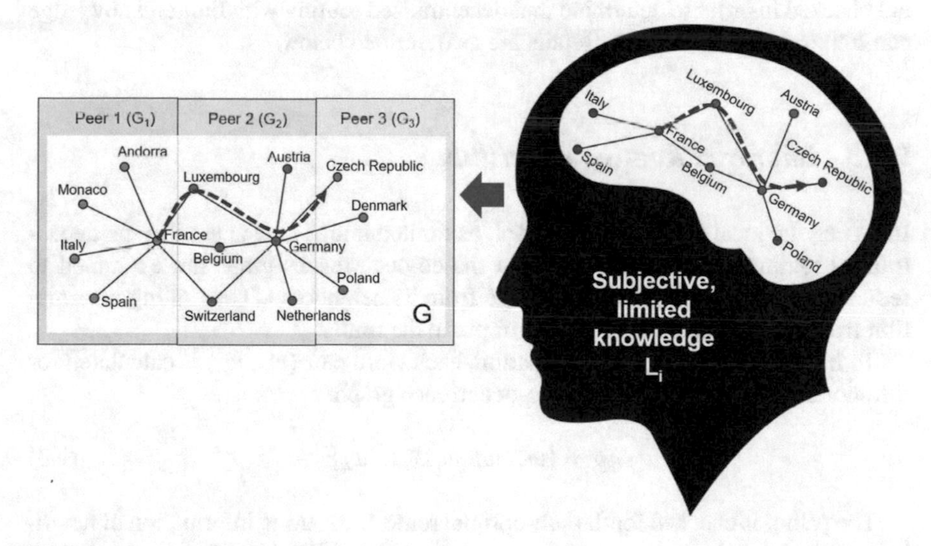

Fig. 9.2 The idea of the local routing method

9.2.2 Routing Analysis on the Distributed Co-Occurrence Graph

In each document, a set of word co-occurrences is inserted into the local and global co-occurrence graphs, a weight function $g((w_a, w_b))$ demonstrates how significant the respective co-occurrence is in a document. The value of $g((w_a, w_b))$ displayed w_a and w_b build undirected co-occurrence graphs will be focused. Dice's coefficient [8] has been used to define the significance measures, the distance d (cost) of the co-occurrence words w_a and w_b can be obtained from the following equation;

$$d(w_i, w_j) = \frac{1}{g((w_a, w_b))} \tag{9.1}$$

The route calculation for all possible connection of source-destination words as pairs (w_a, w_b), when $w_a \in W$ and $w_b \in W$ are determined. The number of routes for analysis on the co-occurrences graph is defined by

$$n(n-1)/2, \tag{9.2}$$

where n is the number of words, i.e., if n is 10, the number of routes is 45.

All possible routes of source-destination word pairs are analysed to ensure that the routing with incomplete knowledge is practical. Both local and global co-occurrences graphs are examined, including unconnected word pairs, connected word pairs, hop counts, and routing with dice coefficient and cost. The goals and results of all routing analyses are presented in Sect. 9.3. Furthermore, all source-destination word pairs are checked in order to determine that decentralised routing with limited knowledge can improve route distances. Details are as described below.

9.2.3 Improved Routing Algorithm

In a peer, the local co-occurrence graph as limited knowledge is used in the decentralised routing process. All routes on the co-occurrences graph are examined to reduce the path by using the knowledge from its neighbours. The routing information from its neighbours may help to improve the route.

To improve the routing of information, each word pair (w_a, w_b) is calculated for the shortest path p from the local co-occurrence graph.

$$p = \{w_a, w_i, w_{i+1}, .., w_b\} \tag{9.3}$$

The path p is checked for the sub-optimal route, the routing information of neighbour peers can help to improve this path p as per the following steps:

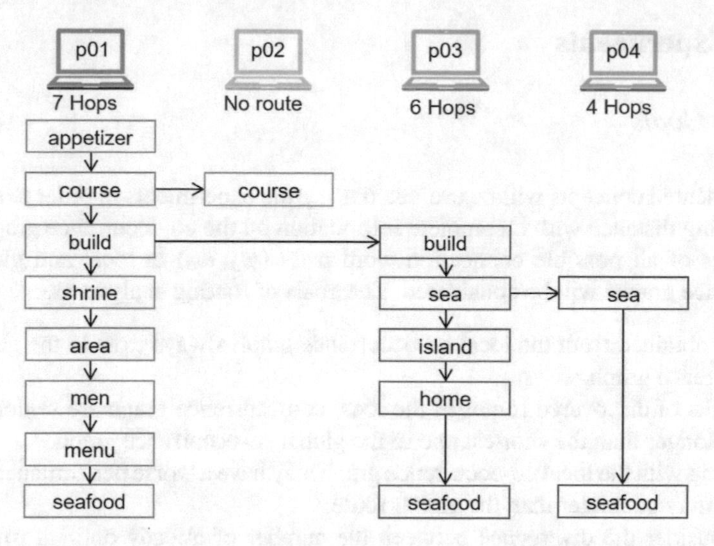

Fig. 9.3 Improving of the path between peers

1. A peer asks for the route information to improve the path p with its neighbours.
2. Every neighbour peer starts to find the shortest path p_n from the first word w_i of path p to word w_b.
3. If there is a shorter path p_n, the path p of words (w_a, w_b) is improved.

$$p = p_n \tag{9.4}$$

   Then, REPEAT step 1.
4. If there is no shorter path p_n, the next word w_{i+1} is processed.
5. If the checked word is the last word of the path p, STOP inhibits this process.

An example of improving a path of a source-destination word pair (appetiser, seafood) is presented in Fig. 9.3. Firstly, peer p01 starts to calculate this word pair's shortest path with its local co-occurrence graph. The route from "appetizer" to "seafood" has seven hops (appetizer => course => build => shrine => area => men => menu => seafood). Then, to improve this path, peer p01 begins to request the routing information from its neighbours. In this example, peer p02 does not have the information to improve this path, whereas peer p03 has the shorter route. Therefore, this path is improved. For the next step, this improved path continues to find the best suitable path. Peer p04 has a better path which can reduce the route from 6 hops to 4 hops. Overall, if there is no better path from any peers, it can be concluded that the path (appetiser => course => build => sea => seafood) is the most suitable path of the source-destination word pair (appetiser, seafood), and the route can be reduced to three hops.

9.3 Experiments

9.3.1 Goals

The presented concepts will be evaluated using the experiments in order to examine the routing distance with incomplete information on the co-occurrence graphs. The distances of all possible connection word pairs (w_a, w_b) in local and global co-occurrence graphs will be considered. The goals of routing analysis are:

- Paths obtained from the local co-occurrence graph always exist in the global co-occurrence graph.
- Lengths of discovered routes in the local co-occurrence graph are seldom found to be longer than the shortest one in the global co-occurrence graph.
- Routing with the local co-occurrence graph may have a worse performance; a local route may be longer than the global route.
- To consider the divergence between the number of already optimal routes and improved routes.
- The decentralised routing with limited knowledge may help to improve the path with sub-optimal routes on the way.

To prove this, as above indicated goals, the experiments are described below.

9.3.2 Experimental Setup

Following a list of documents was inserted, word co-occurrences of those documents were added or updated into local and global co-occurrence graphs. Two datasets were used to build these two co-occurrence graphs as follows:

- First, 100 documents in the travel topic were inserted with 5 peers (20 documents per peer).
- Second, 1000 documents in five topics (200 documents per topic), including arts, cars, computers, leisure, and sports, were inserted with 25 peers (40 documents per peer).

The flow processes of the co-occurrence graphs construction in TheBrain are presented in Fig. 9.4. After a list of documents had been added, the pre-processing and co-occurrences extraction happens on all peers. Consequently, their co-occurrences were concluded, connected, and checked for existence. Then, they were kept in the local and the global co-occurrence graphs. After that, words on the global co-occurrence graph were checked for double-words. Then, peers performed checking for load balancing with their neighbours to attain a good performance. Additionally, peers had the task of connecting several peers through the network using the HTTP protocol to locate words on the global co-occurrence graph.

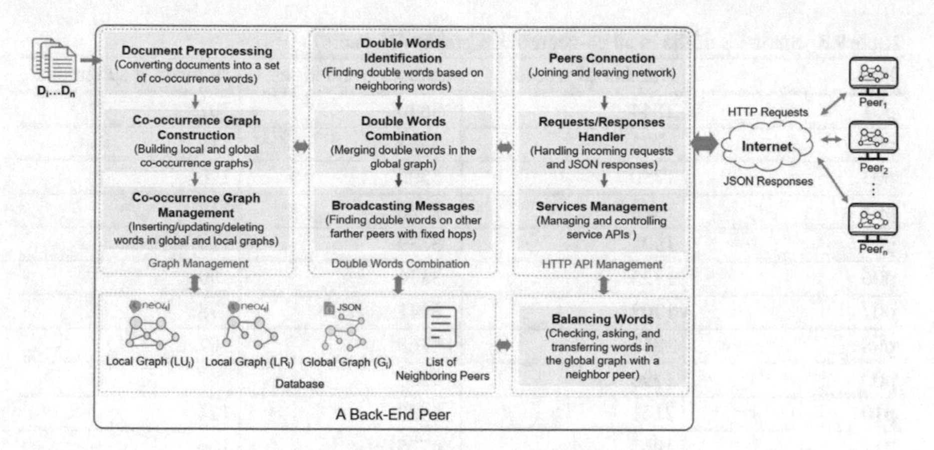

Fig. 9.4 The co-occurrence graphs construction flow processes in TheBrain

Table 9.1 Statistics of the global co-occurrence graph

Peers	Number of words	Number of relations	Number of subgraphs
5	4053	21496	27
25	16371	141475	214

Table 9.2 Statistics of the local co-occurrence graphs (5 Peers)

Peer	Number of words	Number of relations	Number of Subgraphs
p01	1672	5701	28
p02	1254	4500	24
p03	1297	4495	24
p04	811	3058	12
p05	1237	4458	21

The routing analysis begins after the local and global co-occurrence graphs were built from several documents successfully. The number of words, relations, and subgraphs in the global co-occurrence graph with 5 and 25 peers (see Table 9.1) and in the local co-occurrence graphs of all peers (see in Tables 9.2 and 9.3) were analysed according to the goals. The analysis dctails will be described in the following sections.

9.3.3 *Results and Discussion*

The experiments discussed herein present that the proposed concepts successfully compared the routing information between the local and global co-occurrence graphs.

Table 9.3 Statistics of the local co-occurrence graphs (25 Peers)

Peer	Number of words	Number of relations	Number of subgraphs
p01	1844	3640	107
p02	1773	3332	119
p03	1750	3506	106
p04	1528	2626	112
p05	1901	3739	114
p06	1735	3876	99
p07	1703	3941	78
p08	1815	4763	87
p09	1796	4206	87
p10	2141	4528	124
p11	1965	4725	100
p12	1990	23056	83
p13	1921	4171	102
p14	1686	3280	99
p15	1849	3562	146
p16	2099	6133	99
p17	1944	8604	87
p18	2268	12407	96
p19	1919	4601	119
p20	1813	3759	107
p21	1938	3984	113
p22	1909	3538	132
p23	1595	8387	88
p24	1896	11146	78
p25	2476	11561	59

Because every peer has incomplete information, the local routing information was used in the distributed co-occurrence graph. Therefore, the local routing information necessitated verifying that optimal routes or additional routing information from neighbouring peers were required. Besides, the number of already optimal routes and the routes that can be improved have been checked. The experiment results of the routing analysis are presented in Figs. 9.5, 9.6, 9.7, 9.8 and 9.9.

9.3.3.1 Exp. 1: Unconnected Word Pairs

To ensure that the routing mechanism on the distributed co-occurrence graph with incomplete knowledge was suitable to be applied. Firstly, the number of unconnected word pairs (w_a, w_b) in the local co-occurrence graphs were examined. Figure 9.5

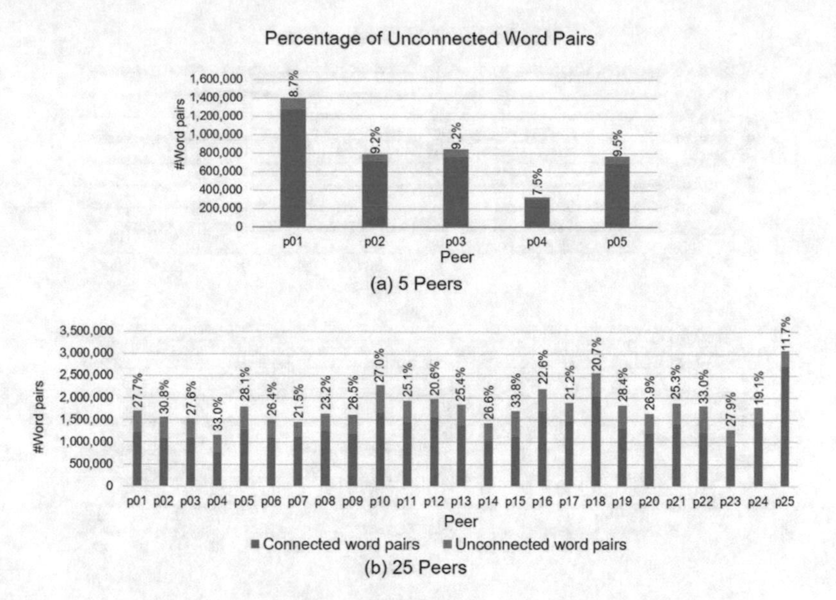

Fig. 9.5 Number of connected and unconnected words in the local co-occurrence graphs

presents the percentage of connected and unconnected word pairs in the local co-occurrence graphs. In 5 peers (see Fig. 9.5a), the percentage of connected word pairs in the local co-occurrence graph was higher than 90%, nearly 10% of unconnected word pairs had a connection in the global co-occurrence graph at 75%. In 25 peers (see Fig. 9.5b), the percentage of connected word pairs in the local co-occurrence graph was approximately 75%. About 25% of unconnected word pairs had a connection in the global co-occurrence graph at 92%. To conclude, the number of unconnected word pairs was minimal. Most of the words have a path to each other. Furthermore, most of the unconnected word pairs in the local co-occurrence graph were the connected word pairs in the global co-occurrence graph.

9.3.3.2 Exp. 2: Hop Counts of Connected Words

The second experiment was aimed to consider the hop counts of all connected word pairs (w_a, w_b) in local and global co-occurrence graphs. The most suitable routing hop was selected to transfer a data packet from source to destination nodes in the routing process. First of all, the hop counts of all connected word pairs in the local co-occurrence graph were considered. In this experiment, most of the routing hops with 5 and 25 peers were between 3 and 6 hops (see Fig. 9.6). The minimal hop count was 1, while the maximal hop counts were 9 (5 peers) and 15 (25 peers). Besides, in the local routes, the average hop counts were 3.64 (5 peers) and 4.66 (25 peers) whereas, in the global routes, the average distances were 3.16 (5 peers) and 3.01 hops (25 peers). Overall, the routing hop in the global co-occurrence graph was

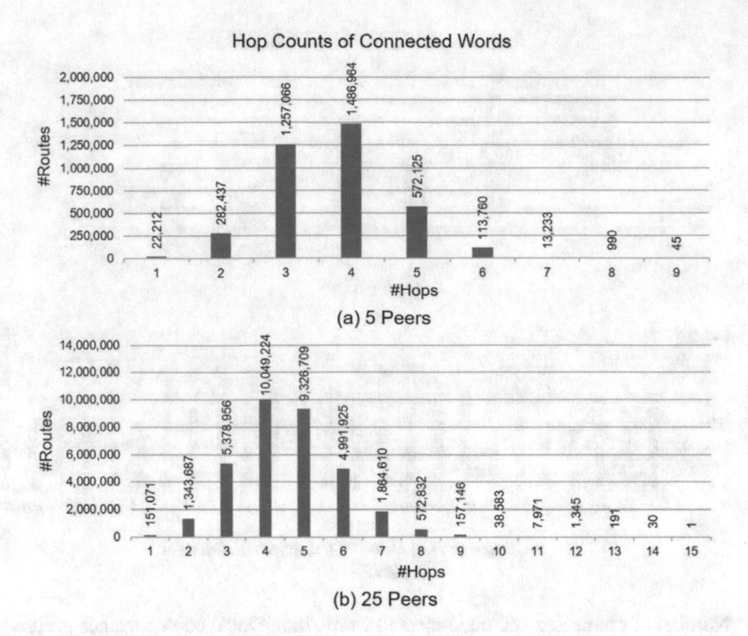

Fig. 9.6 Hop counts of connected words

shorter than the local co-occurrence graph. These indicated that in the distributed co-occurrence graphs, the routing hop could be reduced. The smallest number of routing hops was helpful for data transmission in the network.

9.3.3.3 Exp. 3: The Comparison of Hop Counts Between Local and Global Co-Occurrence Graphs

The comparison of hop counts between local and global co-occurrence graphs was examined in this experiment. The result shows that, in 5 peers, the most compared hop counts were equal (0 hop), and one hop of the global co-occurrence graph was shorter than in the local co-occurrence graph. While in 25 peers, routing hops in the global co-occurrence graph were mainly shorter than in the local co-occurrence graph by one or two hops (see Fig. 9.7). Overall, the lengths of the compared routes between local and global co-occurrence graphs were not greatly different.

9.3.3.4 Exp. 4: Improved Routes in Local Co-occurrence Graphs with Neighbor Peers

In this experiment, all source-destination word pairs in the local co-occurrence graph were checked to determine that the routing could be reduced by using the limited knowledge from its neighbours. Figure 9.8 shows that most of the routes on the local

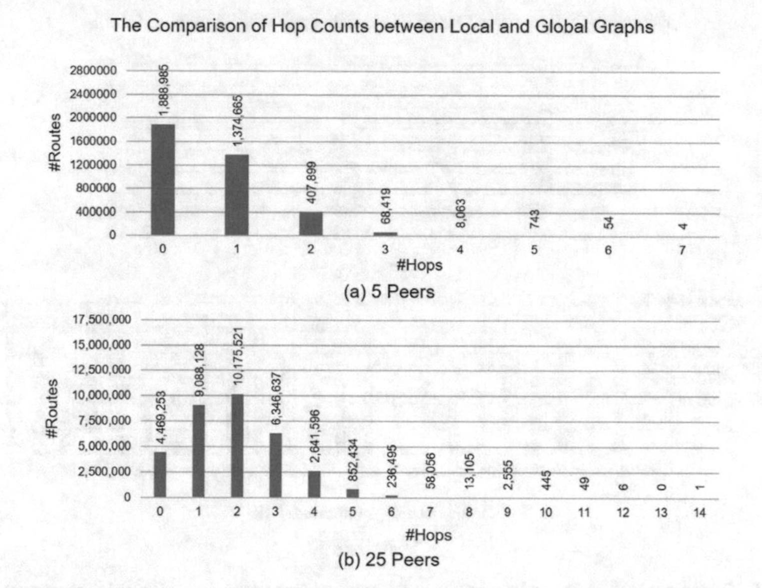

Fig. 9.7 The comparison of hop counts between local and global co-occurrence graphs

co-occurrence graph were already the optimal ones. The percentage of hop counts were minimally decreased at about 3% (5 peers) and 10% (25 peers), which minimally improved the number of hops at 1.2 (5 peers) and 1.5 (25 peers). In conclusion, this indicated that decentralised routing with limited knowledge could reduce the path with the sub-optimal routes.

9.3.3.5 Exp. 5: Routing with Dice Coefficient and Cost

The last experiment demonstrated the number of routes with cost in the local and global co-occurrence graphs. In Fig. 9.9, in 5 peers, the total cost average in the local and global co-occurrence graphs were 69.50 and 90.93 sequentially. Similarly, in 25 peers, the total average cost in the local and global co-occurrence graphs were 66.08 and 169.66, respectively. However, contrary to routing with hops, the paths within the global co-occurrence graph were equal or shorter than in the local co-occurrence graph. Moreover, in the global co-occurrence graph, the average was increased with the routing with cost when more peers were added. The hop count average decreased compared with the local co-occurrence graphs at 0.05 (5 peers) and 0.06 hops (25 peers). From these results, it can be concluded that the routing of individual connections using cost has a slight significance in improving the routing information.

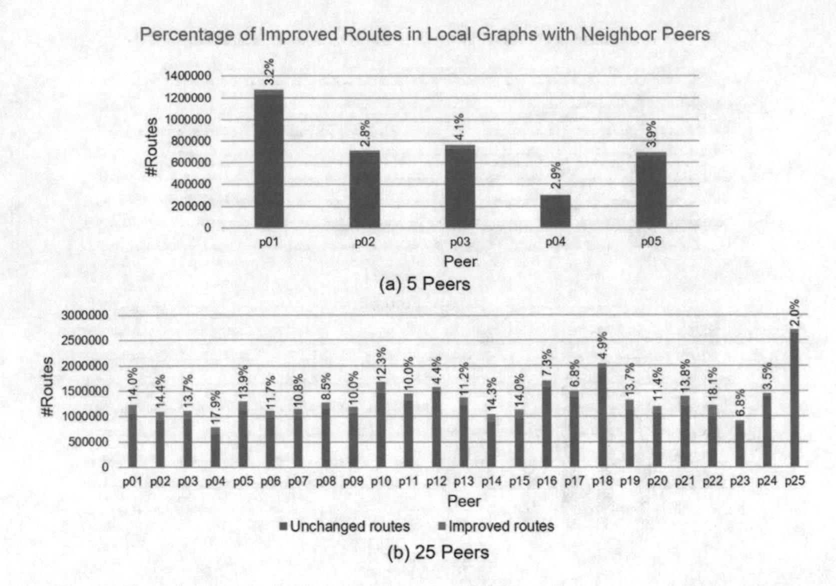

Fig. 9.8 Percentage of improved routes in local co-occurrence graphs with neighbor peers

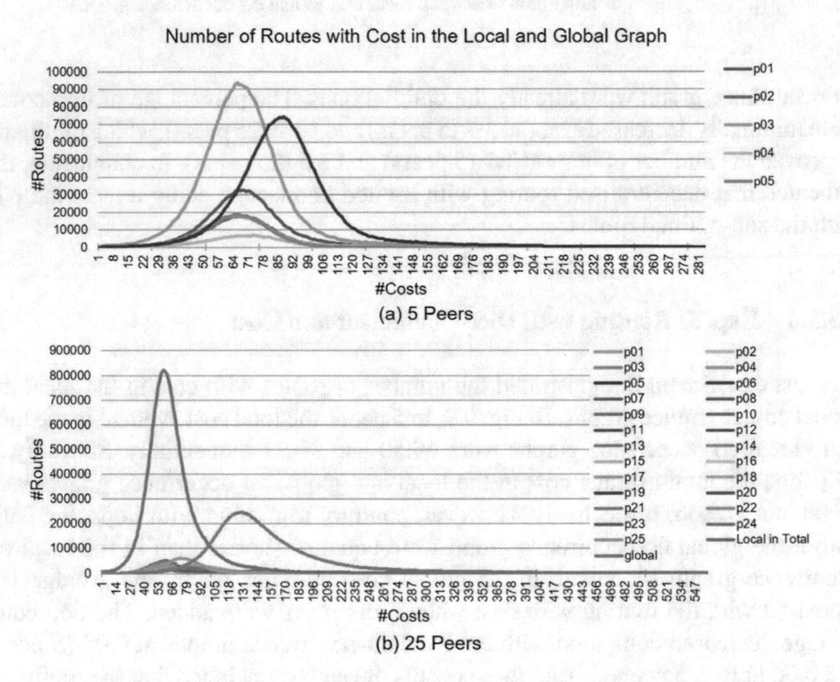

Fig. 9.9 Number of routes with cost in the local and global co-occurrence graph

9.4 Conclusion

The routing on the distributed co-occurrence graph has been analysed using "hop" and "cost" metrics. In comparing routing using the hop, one could observe that the routes in the global co-occurrence graph were shorter than in the local co-occurrence graph with increasing peer numbers. The costs sharply increased when more peers were added. Moreover, the routing hops in the global co-occurrence graph were shorter than in the local co-occurrence graph. The percentage of hop counts were minimally decreased at about 3% (5 peers) and 10% (25 peers), which minimally improved the number of hops at 1.2 (5 peers) and 1.5 (25 peers). Furthermore, the routing with the local co-occurrence graph could help to improve the path from other sub-optimal routes. Overall, the decentralised routing with incomplete knowledge could increase the performance in determining suitable paths. In addition, the proposed routing mechanism can be applied to (decentralised) search engines.

References

1. Stoffel, E.-P.: A research framework for graph theory in routing applications. Master's thesis, Institute of Informatics, LMU, Munich, November (2005)
2. Sanders, P., Schultes, D.: Engineering fast route planning algorithm. In: Experimental Algorithms of 6th International Workshop on Experimental and Efficient Algorithms, pp. 23–36. Rome, Italy (2007) Springer
3. Yang, T., Gerasoulis, A.: Web Search Engines: Practice and Experience, pp. 1367–1390. CRC Press (2014)
4. Dijkstra, E.W.: A note on two problems in connexion with graphs. Numerische Mathematik **1**, 269–271 (1959)
5. Delling, D., Pajor, T., Wagner, D.: Accelerating multi-modal route planning by access-nodes. In: Proceedings of 17th Annual European Symposium, pp. 587–598. Copenhagen, Denmark, Springer (2009)
6. Kubek, M., Unger, H., Dusik, J.: Correlating words—approaches and applications. In: Proceedings of 16th International Conference on Computer Analysis of Images and Patterns, pp. 27–38 Valletta, Malta, Springer (2015)
7. Kubek, M., Unger, H.: The webengine—a fully integrated, decentralised web search engine. In:Proceedings of the 2nd International Conference on Natural Language Processing and Information Retrieval, pp. 26–31. Bangkok, Thailand (2018). Association for Computing Machinery
8. Dice, L.R.: Measures of the amount of ecologic association between species. Ecology **26**, 297–302 (1945)

Chapter 10
Sequence Identification with Trees and Co-Occurrence Graphs

Maximilian Knoll and Herwig Unger

10.1 Introduction

Sequence learning is an important concept in learning [1] and is utilized in a wide range of disciplines; applications include sequence identification and prediction. Especially for the navigation of the internet (search engines, text-mining) and interactive applications as chat bots, extraction of frequent and relevant sequences is crucial.

For sequence identification, a wide range of techniques has been described, with specific (dis-)advantages depending on the task: recurrent neural networks, e.g., exhibit a considerable degree of oblivion over time [2]. Alternative approaches include SPADE [9], for an overview see also [10].

In natural language processing (NLP), collections of texts (corpus), are frequently analyzed sentence wise. Single sentences are usually further processed to retain only word-stems of nouns and names [3]. Co-occurrence graphs enable identification of frequently co-occurring words, which represent the strength of semantic connections. Each node corresponds to a word, edges between nodes contain their co-occurrence frequency [3]. Such analyses aim to extract relevant words describing the underlying text(s).

We evaluate a tree based approach for the identification of sequences in co-occurrence graphs. A tree is appended to each node of the co-occurrence graph and is filled over time, containing sequences starting with the respective node (root). We longitudinally assess the impact of tree parameters height, degree and varying replacement strategies on the ability to identify (meaningful) sequences.

H. Unger and M. Kubek, *The Autonomous Web*, Studies in Big Data 101,
https://doi.org/10.1007/978-3-030-90936-9_10

10.2 Data and Methods

10.2.1 Text Corpus

A news article corpus consisting of 3723 news articles included in the Hagen NLP toolbox eclipse project [4] (Sueddeutsche Zeitung, 2015-09-01 to 2015-11-04) was used to create a co-occurrence graph using the toolbox with default parameters (n = 72993 nodes).

10.2.2 Trees

To each of the co-occurrence graphs nodes', a tree was appended, with the respective node being the root of the tree. Each tree has a height h and a degree d. The total number of nodes for a given tree is $\sum_{i=1}^{h} d^{i-1}$. Child node indices i range from $i_{from} = d \cdot i_{parent}$ to $i_{to} = d \cdot (i_{parent} + 1) - 1$ for $i_{root} = 1$ (see Fig. 10.1).

Each node corresponds to an item/word in a given sequence, the leading edge (weight w) indicates the frequency at which this node was reached. The time-point t of last access/creation is stored in each node. Indices of nodes are stored as `BigInteger`, tupels of indices and nodes (`Node` class instances) in a `HashMap`.

For retrieval of sequences, the tree is traversed starting from the root, usually with a weight cutoff to select only frequent associations. For the tree shown in Fig. 10.1, all sequences with weight ≥ 2 are *ea*, *eab*, *eabc*, *ead* and *ef*.

10.2.3 Tree Assembly

$G = (V, E)$ denotes the co-occurrence graph with nodes V and edges E. Let S be a set of sequences starting with word r. If a V_j exists with name r, a tree is appended to V_j such that V_j is the root of the resulting tree with index 1.

Let $s \in S$ be a sequence with $s := (q_1, q_2, ..., q_n)$ of words q_k. q_1 is the name of a V_j, $q_1 := r$. Processing each word q_k from a sequence s from left to right for tree assembly is performed as follows (i is the node index for storage, see Sect. 10.2.2):

1. Compute child node indices $[i_{from}, i_{to}]$ for the node index which contains the node with name q_k.
2. Check if i_{from} is already occupied, if not, create a new node V^* with name q_{k+1}. Continue with (1) with q_{k+1}.
3. If a node exists at i_{from}, and the name of this node is q_{k+1}, update this node (last access, increase weight). Proceed with (1) with word q_{k+1}.
4. If a node exists but the node name does not equal q_{k+1}, increment the position to test (i_{from}) until $i_{from} + 1 \leq i_{to}$ and go to (2).

Fig. 10.1 Tree for sequence identification (**a**) and its storage (**b**). Superscript: index, edges: weights. Gray: nodes with weights ≥ 2

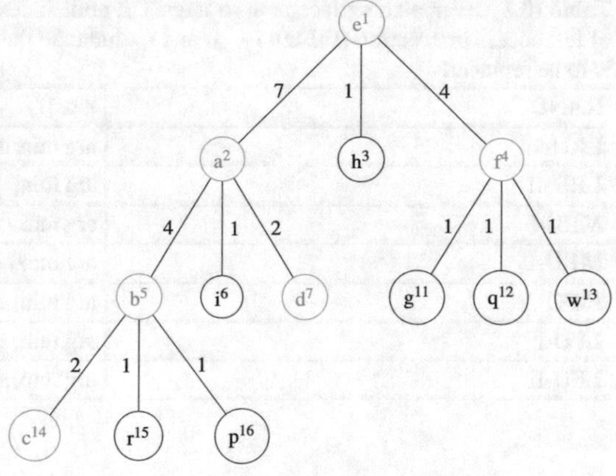

(a) Tree with node indices.

index	0	1	2	3	4	5	6	7	8	9	10	11
node		e	a	h	f	b	i	d				g

(b) Storage of tree.

5. If no insert / update was possible, replace a node with index $z \in [i_{from}, i_{to}]$ based on the selected replacement strategy. Remove all children of node with index z. If the sequence s contains an unprocessed word, proceed with (1).

10.2.4 Replacement Strategies

If adding a sequence to a tree encounters a filled level (child indices ranging from i_{from} to i_{to} for a given node index, a subtree / node is replaced. Utilized replacement strategies are shown in Table 10.1. For the random replacement strategy (RAND), a fixed seed was used to obtain reproducible results.

10.2.5 Sequence Processing

Lexicographically ordered files (by filename) were processed per sentence.

1. Convert next sentence to sequence of word-stems and names with Hagen NLP toolbox. Proceed with (2).
2. Test if first word of the sequence is contained in the co-occurrence graph, if yes (3), else (4).

Table 10.1 Evaluated replacement strategies. i: node index, w weight of leading edge, t timepoint of last access or creation (FIFO). i_{from} and i_{to} indicate the range of child indices from which one is to be replaced

RAND	$r \in [i_{from}, i_{to}]$
LRU-I	$\arg\min_i t_i \cdot w_i$
LRU-II	$\arg\min_i \frac{t_i}{w_i}$
MRU-I	$\arg\max_i t_i \cdot w_i$
MRU-II	$\arg\max_i \frac{t_i}{w_i}$
FIFO	$\arg\min_i t_i$
LFU-I	$\arg\min_i w_i$
LFU-II	$\arg\min_i t_i$

3. Add sequence to tree: follow the sequence from the root of the tree and insert or update node(s) in subtrees, replace if necessary, truncate sequence if necessary (length >tree height). See also Sect. 10.2.3. Proceed with (4).
4. Remove first word from sequence, if sequence length >1 (2), else (1).

Processing time and tree characteristics (mean and maximum number truncations and replacements and approximate number of utilized nodes[1]) were evaluated every 50th file by non-linear model based regression (see Sect. 10.2.7).

10.2.6 Sequence Evaluation

Sequences were extracted from trees every 50th file (see Sect. 10.2.2 and Fig. 10.1), using weight cutoffs of 1 to 7. Sequences were manually scored with a semi-quantitative score ranging from 0 (no meaning) to 3 (clear meaning). Dimensionality reduction for meaningful longitudinal sequence identification profiles was performed using a specific tSNE implementation [5].

10.2.7 Model Based Evaluation

To evaluate time requirements, tree characteristics and as (meaningful) identified sequences over the process of news article analysis, a range of non-linear models (sigmoidal curves: logistic, Gompertz and modified Gompertz functions; asymptotic regression models and others, see [6]) was fitted to transformed data. Both dependent and independent data were log or $identity$ transformed, residual standard error (RSE) distributions were used to select transformations, and rank analysis of RSEs

[1] $\frac{1}{10}\sum_{i=1}^{10} d^{l_i}$, d: degree, l_i: length of random sequence i with weight ≥ 1, $1 \leq i \leq 10$.

(min, mean, median, max) yielding lowest values were used to select appropriate models. Plateau parameters as e.g. model parameter d (upper asymptote value) for sigmoidal curve models were evaluated to characterize the impact of varied parameters using linear models.

10.2.8 Statistical Analysis

Statistical analyses were performed in R [7]. The *drc* package [8] was used for nonlinear model fits utilizing functions from the *aomisc* package [6]. Residual standard errors were used for comparative model evaluation (selection of appropriate nonlinear model functions).

Likelihood ratio tests between null and full models were used to calculate global model p-values. Differences between two groups are reported as Wald-type p-values from linear model analyses. χ^2 tests were used to test associations between categorical data. Significance level α was set to 0.05 (two-sided). log denotes the natural logarithm.

10.3 Results

10.3.1 Time Requirements and Tree Metrics

A qualitative overview of analysis results of longitudinal data using models for time requirements and tree characteristics (average and maximum number of truncations, replacements and approximate number of nodes per tree) is shown in Table 10.2, evaluating upper asymptote parameter d or plateau parameter.

Time requirements increased non-monotonously with height and tended to decrease with degree, no difference between replacement strategies was observed.

Approximate number of nodes increased with degree. Truncations decreased with height and degree (not for maximum truncations for the latter). Replacement strategy did not yield significant differences. Maximum replacements per tree showed increasing values for higher degrees and decreasing values for average replacements. Tree heights did not show differences, low numbers of maximal replacements were observed for FIFO and LFU-II.

10.3.2 Sequence Identification

Longitudinal data of sums of sequences with weights ≥ 7 detected with the present approach are shown in Fig. 10.2. Smaller degrees yielded lower numbers of

Table 10.2 Effect of analysis parameter variation on higher asymptote model parameter *d* or plateau parameter. Arrows indicate directions of median parameter value changes with increasing height (reference 5) and degree (reference 3) or between replacement strategies (reference: RAND), model p-value, likelihood ratio test. ↑, ↓: monotonous increase/decrease, non-monotonous significant differences otherwise

Parametr	Height↑	Degree↑	Replacement	
			↓⁰	↓⁰
Time requirement+	↗	(↘)	–	–
Mean (approx) nodes?	–	↑	–	–
Mean truncations+	↓	↘	–	–
Max truncations+	↘	–	–	–
Mean replacements+	–	↘	–	–
Max replacements+	–	↑	RAND, MRU-II	LFU-II, FIFO
Sum sequences+				
Weight 1	↗	↑	LFU-I, LRU-I	LRU-II, FIFO
Weight 2	–	↷	RAND, MRU-II	LRU-I
Weight 3	↗	–	FIFO, LFU-II	LFU-I
Weight 4	–	↗	FIFO, LRU-II	LFU-I
Weight 5	–	↘	FIFO	MRU-I
Weight 6	–	↘	FIFO	MRU-I
Weight 7	–	↘	FIFO	MRU-I
max sequences+				
Weight 1	↑	↑	LFU-I	MRU-I, FIFO
Weight 2	↑	↑	LFU-I, LRU-I	LRU-II, MRU-I, FIFO, LFU-II
Weight 3	↑	↑	LFU-I, LRU-I	LRU-II
Weight 4	↑	↑	LFU-I	FIFO, LRU-II
Weight 5	↑	↑	LFU-I	FIFO, LRU-II
Weight 6	↑	↑	LFU-I	FIFO, LRU-II
Weight 7	↑	–	LFU-I	FIFO, LRU-II

?: asymptotic regression model, plateau parameter
+: modified Gompertz model (4 parameters), upper asymptote parameter
0: vs RAND

sequences, replacements strategy showed differences: highest values were observed for LFU-I, LRU-I, MRU-II and RAND; FIFO and LRU-II yielded lowest values.

Qualitative data for model based analyses are shown in Table 10.2. Results were more heterogeneous for sum of sequences analyses for increasing weight thresholds as compared to maximum numbers of sequences. For the latter, a trend of increasing values with increasing height and degree was observed in all but one case (weight $\geq$ 7 and degree). For sum of sequences, a positive association with degree and height was observed for weight ≥ 1, for weights >4, however, height showed no significant association and a trend towards a decrease was observed for degree.

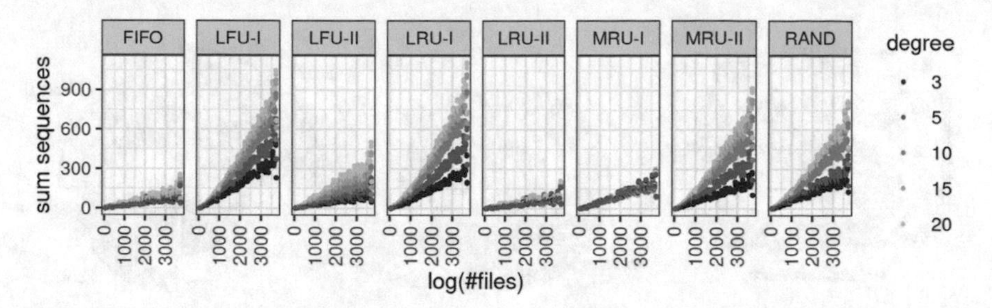

Fig. 10.2 Sum of detected sequences for weight ≥ 7

For larger weights, LFU-I yielded higher values for maximum numbers of sequences, FIFO for sum of sequences. FIFO and LRU-II, however, showed for maximum number of sequences and higher weight cutoffs significantly lower values as compared to RAND.

10.3.3 Sequence Evaluation

From the n=7442 unique sequences with weight ≥ 4 (Fig. 10.3a), 38% (n=2933) had lengths ≥ 3 (Fig. 10.3b). Nearly one fourth (21%) of the latter was manually scored as meaningful (score 3, Fig. 10.3c), 29% were classified as non-meaningful (score 0).

Sequences with score 3 were further analyzed. Detected overlaps with recognized sequences over the curse of analysis for different evaluated parameters combinations are show in Fig. 10.3d. Differences between replacement strategies decreased with higher degrees. For lower degrees, two qualitatively different groups of longitudinal profiles could be already distinguished (see below for detailed evaluation).

Model based analyses (Gompertz, with lower limit at 0) of data shown in Fig. 10.3d is depicted in Fig. 10.3e–g. Height did not show significant differences. Highest values were observed for degree 3 (Fig. 10.3g). Significant differences between replacement strategies were observed: smallest median values for MRU-I, highest for LRU-II and FIFO (Fig. 10.3e).

10.3.4 Similar Performing Parameter Combinations

As a multitude of parameter combinations was evaluated for meaningful sequence detection, similarity w.r.t. outcome (longitudinal profiles of identified meaningful sequences, see Fig. 10.3d) between combinations was evaluated. Dimensionality reduction of $V = (\boldsymbol{v}_{z_1}|\boldsymbol{v}_{z_2}|...)$ with $\boldsymbol{v}_i$ containing overlaps with all meaningful

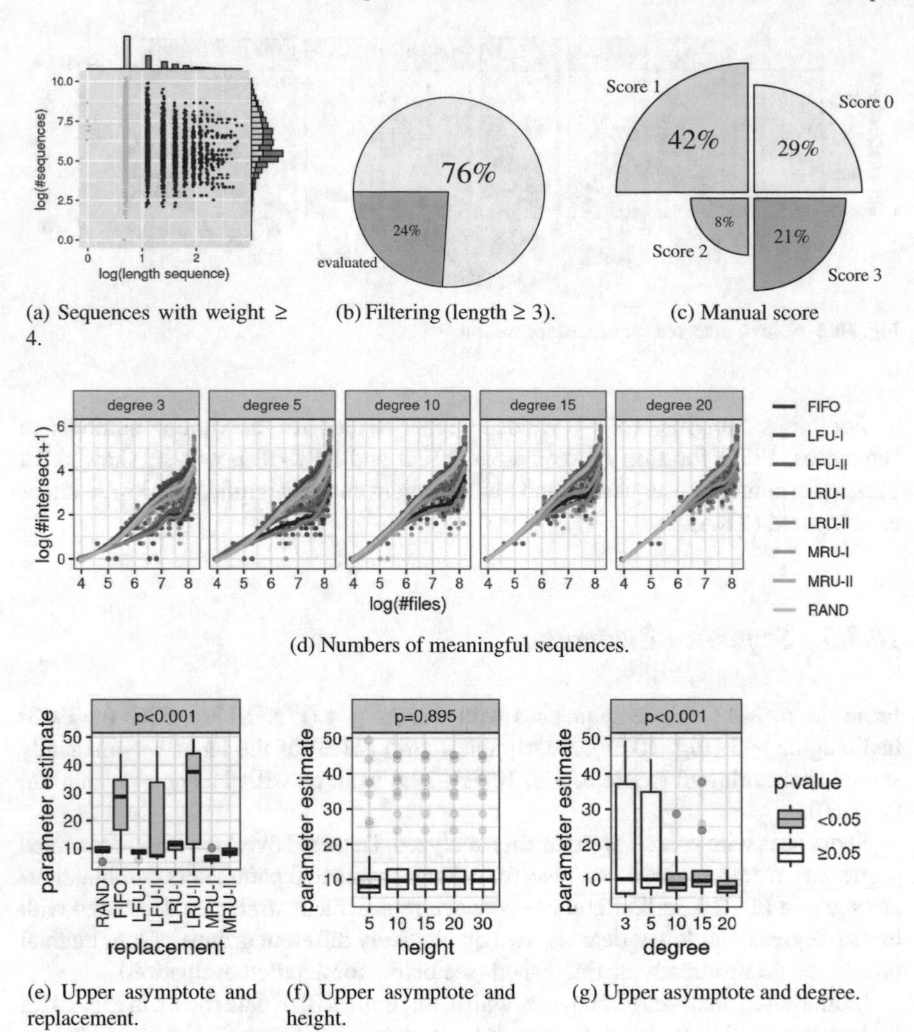

(a) Sequences with weight ≥ 4. (b) Filtering (length ≥ 3). (c) Manual score

(d) Numbers of meaningful sequences.

(e) Upper asymptote and replacement. (f) Upper asymptote and height. (g) Upper asymptote and degree.

Fig. 10.3 Distribution of all sequences with weight ≥ 4 and meaningful sequences. d: loess fit. e-g: reference: RAND, height 5, degree 3. Linear model p-values

sequences for a single combination z ($z :=< h, d, r; w, l >$, h height, d degree, r replacement strategy, w weight cutoff, l length cutoff) for each evaluated timepoint (number of processed files) was performed (t-SNE, Fig. 10.4a)

Two main clusters (group GRP1 and GRP2) of combinations yielding similar results could be identified (silhouette test for k-means clustering, data not shown). Visual inspection revealed no differential distribution of height (Fig. 10.4), higher degrees were associated with larger tSNE-1 values for a given tSNE-2 value (Fig. 10.4). Replacement strategies tended to cluster together.

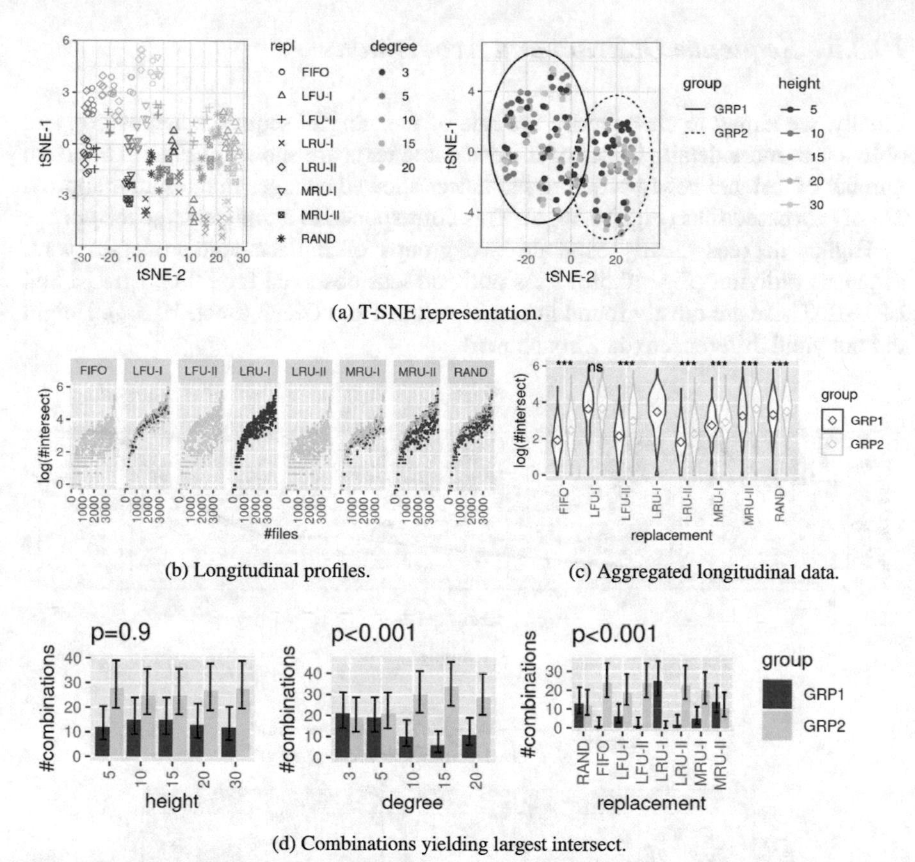

(a) T-SNE representation.

(b) Longitudinal profiles.

(c) Aggregated longitudinal data.

(d) Combinations yielding largest intersect.

Fig. 10.4 Identification and characterization of parameter combinations yielding similar outcomes. c: linear model p-values, ***: $p < 0.001$, *: $p < 0.05$, ns: $p > 0.05$, diamond: median, e: chi^2 tests, Wilson 95% confidence intervals

Longitudinal meaningful sequence profiles colored by group are depicted in Fig. 10.4b, collapsed data over time in Fig. 10.4c. Variation was highest for FIFO, LFU-II and LRU-II. Higher values were observed for GRP2 except for LFU-I. LRU-I was exclusive to GRP1.

Distributions of height, degree and replacement parameters in combinations yielding the highest number of meaningful sequences were assessed for differences between groups GRP1/GRP2 (Fig. 10.4d). For height, no difference was observed, degree showed significant differences between groups for higher degrees (maximum for GRP2, minimum for GRP1 at degree 15). FIFO, LFU-II, LRU-II and MRU-I were more frequently present in GRP2 with high values.

10.3.5 Sequence Oblivion and Acquisition

Finally, we aimed to characterize the rate of meaningful sequence acquisition and oblivion in more detail. Topics of the evaluated texts are shown in Fig. 10.5a. Both number of lost and newly acquired sequences showed changes in rates after approx. 2/3 of processed files (Fig. 10.5b, c). This corresponds to a topic change to *sport*.

Higher degrees clearly separate two groups of replacement strategies w.r.t. sequence oblivion (Fig. 10.5b). Less oblivion was observed for FIFO, MRU-I and LRU-II. These are mostly found in combinations from GRP2 (Sect. 10.3.4). Height did not yield differences (data not shown).

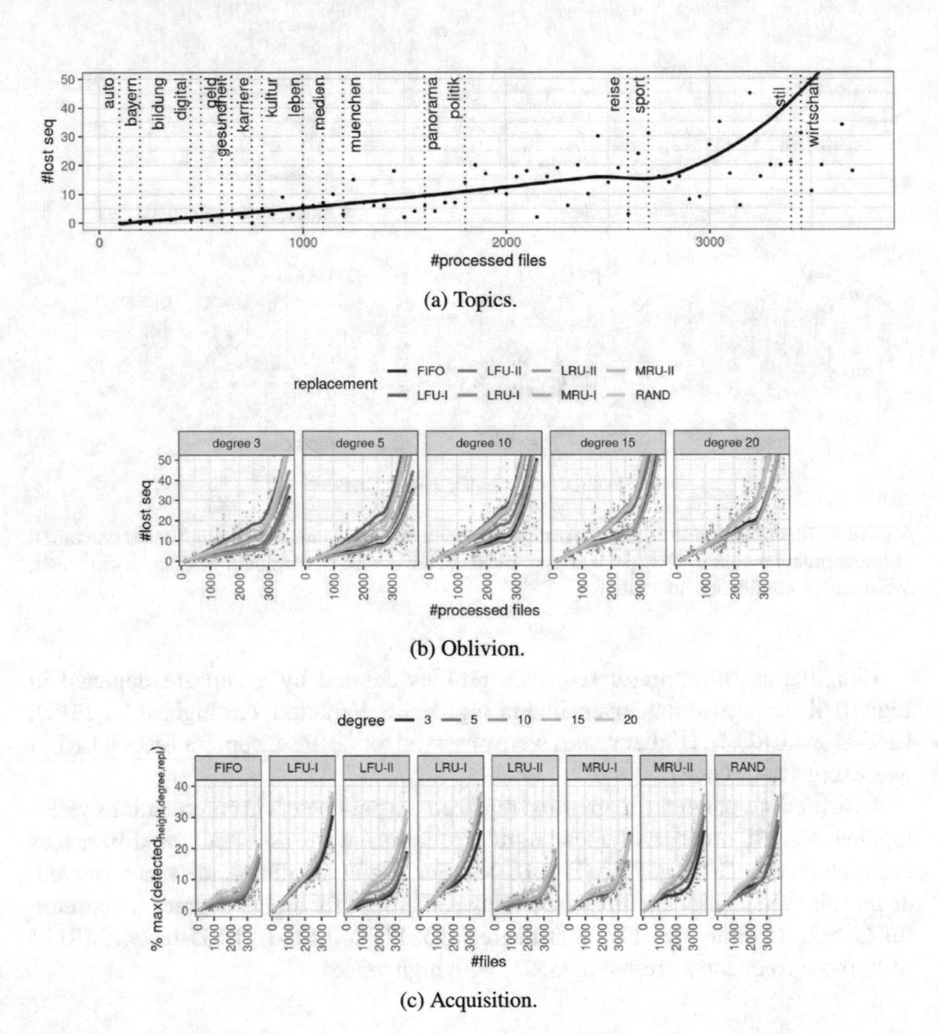

(a) Topics.

(b) Oblivion.

(c) Acquisition.

Fig. 10.5 Oblivion and acquisition of sequences between consecutive evaluation timepoints and topics of analyzed texts. Loess fits. a: degree 3, height 10, RAND replacement

Acquisition of sequences, shown as fraction of identified meaningful sequences from maximum number of meaningful sequences for a given combination of height, degree and replacement strategy showed differences mostly affected by replacement strategy and to a lesser extent by degree (Fig. 10.5c). Height did not show differences (data not shown). Again, two main groups of curves could be distinguished: LFU-I, LRU-I, MRU-II, RAND and LRU-I (corresponding to GRP1, Sect. 10.3.4). These combinations seem to be less influenced by the topic of evaluated texts, but yielded lower numbers of meaningful sequences overall.

10.4 Conclusion

We evaluated the feasibility of using trees in co-occurrence graphs for the identification of sequences in texts. The impact of tree parameters height, degree and replacement strategy on total numbers and numbers of meaningful sequences (manually scores) was assessed. Longitudinal data (over the curse of processing data from a corpus in fixed order) was evaluated directly and by fitting to a variety of non-linear models. Parameter values were evaluated for differences. Two main groups of combinations yielding similar results (longitudinal profiles) were observed. Meaningful sequences were mostly short, thus tree heights $\leq$ 10–15 might be sufficient (higher time consumption for higher trees). Degree 15 and replacement strategy LFU-I yielded robust results. Oblivion of meaningful sequences depended on degree, replacement strategy (low for LRU-II and FIFO) and text topic. Further studies evaluating the effect of text order (topic) on results and validation in independent datasets are warranted.

References

1. Hawkins, J., Blakeslee, S.: On Intelligence: How a New Understanding of the Brain Will Lead to the Creation of Truly Intelligent Machines. Henry Holt and Company, New York (2005)
2. Sun, R., Giles, C.: Sequence learning: Paradigms, Algorithms, and Applications. Springer, Berlin (2000)
3. Kubek, M.: Concepts and Methods for a Librarian of the Web. Cham, Springer Nature Switzerland AG (2020)
4. Kubek, M.: The Hagen NLPToolbox (2019)
5. Linderman, G.C., Rach, M., Hoskins, J.G., Steineberger, S., Kluger, Y.: Fast interpolation-based t-SNE for improved visualization of single-cell RNA-seq data. Nat. Methods (2019)
6. Onofri, A.: Aomisc: Statistical methods for the agricultural sciences, R package version 0.62 (2020)
7. R Core Team: R: A language and environment for statistical computing (2018)
8. Ritz, C., Baty, F., Streibig, J.C., Gerhard, D.: Dose-response analysis using R. PLOS One **10**(12), e0146021 (2015)

9. Zaki, M.J.: SPADE: An efficient algorithm for mining frequent sequences. Mach. Learn. **42**, 31–60 (2001)
10. Mukherjee, S., Rajkumar, R.: Frequent item set, sequential pattern mining and sequence prediction: structures and algorithms. In: International Conference on Intelligent Computing and Smart Communication (2019)

Chapter 11
Oblivion in Time-Dependent Information Management

Erik Deussen, Marius Immelnkämper, and Herwig Unger

11.1 Motivation and Background

Every language changes more or less rapidly over a period of time, reflecting current developments in society or events that occur. Therefore, a person born at the beginning of the 18th century would only with difficulty understand his own mother tongue today and vice versa. Certain regions, especially geographically isolated ones, develop their own dialect for this reason; in this context, language, apart from the communication aspect, also fulfils another task as a group identifier or for the exclusion of non-group members from a certain community. Several factors contribute to language development:

- The already mentioned geographical separation or isolation.
- A historically conditioned change in the meaning or evaluation of the German words (e.g. 'Weib' and 'Fräulein' became uncommon).
- Social change makes certain terms disappear or changes their meaning and use (e.g. terms like the Soviet Union, communism).
- Fashion words and youthful word creations (super, awesome, hyper, in) as well as synthetic words created by advertising to attract attention, which in part also create a new word meaning (the (Nissan) Cube, which in addition to cube now also stands as a designator for a special car and its design).
- The adoption of terms from other languages due to naming, migration, or designation of foreign or unknown things (especially for German Anglicism such as 'Sandwich', 'performen', 'Einchecken', 'business plan', 'News', etc.).
- Words newly created by special events or development (e.g. CoVid-19) or else a marked change in the frequency of use of words (e.g. fever, headache, at the beginning of a flu epidemic).
- Word adaptations forced by changing social values and laws (e.g. the gender language 'StudentInnen', 'Student*innen' or adaptations in spelling due to reforms).

H. Unger and M. Kubek, *The Autonomous Web*, Studies in Big Data 101,
https://doi.org/10.1007/978-3-030-90936-9_11

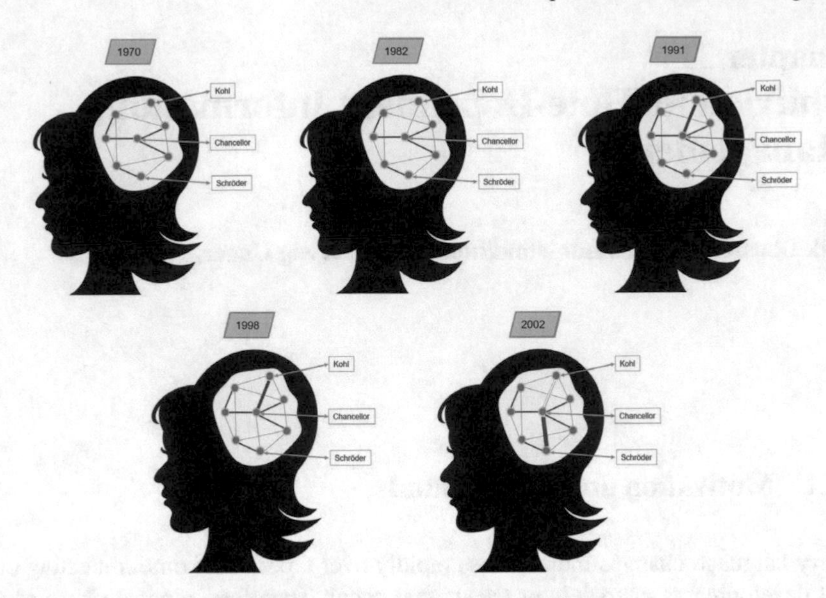

Fig. 11.1 Changes in co-occurrence graph

Those language updates will also result in changes in co-occurrence graphs: graphs established in different periods, so-called diachronic co-occurrence graphs [5], will exhibit separate sets of nodes and edges.

Figure 11.1 shows a respective example. While in 1972 the words Kohl, Chancellor and Schröder were not in co-occurrence, this co-occurrence relationship changed at dedicated points in time and events: first with the election of Kohl as Chancellor, then again through his significant role in the formation of German unity. With his deselection in 1998, the word Chancellor co-occurred more frequently with the name of the new Chancellor (Schröder) and the old association 'Chancellor Kohl' lost significance (comparing edge weights). Only the death of the 'Chancellor of Unity' in 2017 led to a reappearance of the name and the associated co-occurrences for a short period in the news. Of course, those changes are in diachronic corpora are investigated over larger periods only.

The construction of a co-occurrence graph from a natural language using an NLP programming library for text processing (e.g as in [1]) is mostly static in its nature and takes little account of the temporal-dynamic evolution of language, although this seems to be occurring at ever shorter intervals these days.

In [4, 5], Heyer and others investigated for the first time short time changes, so-called weak signals. They figured out that new appearing events are not represented by high frequently used words but by those changing their frequency rank rapidly. The introduced *volatility of words* was a measure for such changes, see also [7].

Last but not least, also clustering and in particular dynamic and hierarchic cluster methods (for an overview see [6]) give an opportunity to a dynamic analysis of maybe changing topics, origins of information or generalisations.

Fig. 11.2 Human learning behaviour

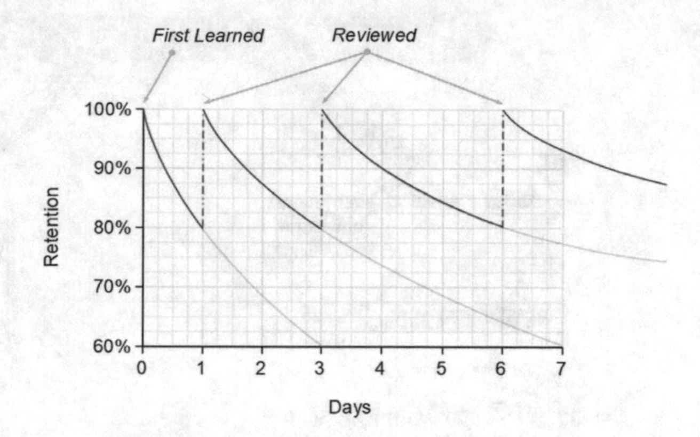

However, in most cases, extensions of the co-occurrence graph have been considered and not the removal of nodes and edges. Nevertheless, this would be the needed process to model the oblivion of outdated and renewed news as well as not well established fake news. Consequently, in the recent contribution, the modelling of oblivion processes shall be considered and the first simulation results will be presented.

In other contexts, i.e. in a learning environment, those processes have been already considered in a macroscopic way [3], as it can be also seen in Fig. 11.2, while hereby, in particular, the role of repetition of knowledge in intervals and the following oblivion with different time constants were considered and played a major role.

11.2 Concepts

11.2.1 Oblivion Processes

The explanations in the above section show that there are occasions in the development of languages requiring the removal of words and their co-occurrences from a co-occurrence graph, while also a significant reduction of the weights of an edge can be used to model the observed processes in reality. The latter is a process, [2] already modelled in a similar manner for his ant hill experiences by the decay of pheromones (and the update of the ant hill's social map) in another context.

Applying a similar process to learning and oblivion and therefore node and co-occurrences of the co-occurrence graph, one can model it for every time step (in our case every reading and processing of a new document) as follows:

1. Learning new words result in an addition of nodes to the co-occurrence graph.
2. Co-occurrences found in the read text result in new edges or weight updates in the co-occurrence graph.

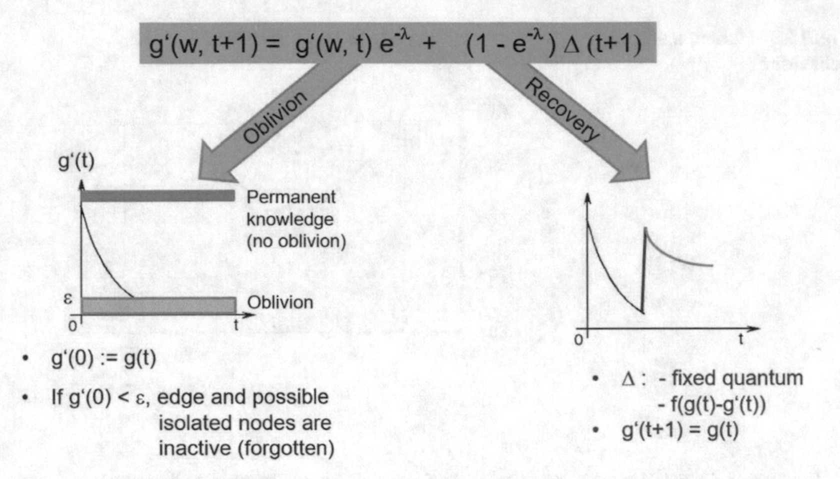

Fig. 11.3 Modelling of the oblivion process

3. Longer not recognised words and co-occurrences will be subject to oblivion, i.e. the weight of the edges in the co-occurrence graph must be reduced. Typically this is an exponential process. Some of those elements may not be included in this process, since they have been so often presented that they can be considered as permanent knowledge.
4. There shall be a recovery process, i.e. an opportunity to increase the weight of edges if a word or a co-occurrence appear again after a long time; as in the human brain, this shall result in some recall effects.
5. Words once known usually do not forget, even if not used anymore. This means isolated nodes in the co-occurrence graph will not be considered anymore but shall not be deleted due to possible edge recovery processes.

Figure 11.3 shows the whole modelling process. It requires the following changes and additional information and processing steps:

1. In the co-occurrence graph $G(t) = (W(t), E(t), g(E, t)$, a second weight function $g'(E, t)$ must first be introduced in parallel to the existing edge weight, in which updated values can be stored independently of the actual frequency of occurrence of a co-occurrence involving forgetting processes.
2. The moment an edge $(w_i, w_j) \in E(t)$ is inserted, where $g'((w_i, w_j), t) = g((w_i, w_j), t)$ is set.
3. After the expiry of a time unit at $t + 1$, an exponential forgetting step is added to the calculations, i.e. $\forall((w_i, w_j), t + 1) \in E(t + 1)$:

$$g'((w_i, w_j), t + 1) := g'((w_i, w_j), t) \cdot e^{-\lambda}.$$

Note: Here we again assume that the processing of a document corresponds to a time step, which means that the corresponding model investigations remain

reproducible. Generally speaking, however, an absolute time scale must be applied to the forgetting processes, in which pauses and processing speed naturally have an influence on forgetting.
Also, the same oblivion rate λ is used for all processes without any difference.
4. It is assumed that there is a certain amount of basic, permanent knowledge that cannot be forgotten. This is knowledge with a very high, significant frequency of the respective co-occurrence, i.e. $g((w_i, w_j), t) > \alpha$; where α is a constant describing this limit. These edges are not subject to the forgetting process described in the previous step, and it further holds that $g'((w_i, w_j), t+1) := g'((w_i, w_j), t) = g((w_i, w_j))$.
5. Depending on whether a co-occurrence appears in the document $D(t+1)$ read at time $(t+1)$ or not, a memory effect is triggered, which leads to an increase in the edge value subject to forgetting $g'((w_i, w_j), t)$, where at most the value of $g((w_i, w_j), t)$ can be reached. Consequently, $\forall((w_i, w_j), t+1) \in E(t+1)$ the edge weights $g'((w_i, w_j), t+1)$ are updated by

$$g'((w_i, w_j), t+1) := g'((w_i, w_j), t) + \Delta$$

whereby Δ results as follows:

- **0**, if the co-occurrence $((w_i, w_j))$ does not appear in $D(t+1)$ and
- $\mathbf{g((w_i, w_j), t+1) - g'((w_i, w_j), t+1)}$ otherwise.

Note that the chosen way of calculating Δ is only one of the possibilities; depending on the application and concrete circumstances, one could also just assign a fixed part of $g((w_i, w_j), t+1) - g'((w_i, w_j), t+1)$ or even assign a fixed quantum of Δ.

Note that there is a significant difference between this and the procedure for ant colonies, for example. By merely setting edges inactive, they are not available for current calculations, but a memory process in the sense of a human 'aha'-effect remains possible. With the changes and additions described, a completely new learning process inspired by human thinking can be modelled, which for the first time includes different approaches and stages of forgetting in addition to the acquisition of knowledge. These will be presented systematically in the following section.

11.2.2 Systematics

According to the calculations shown above and the diagrams in Fig. 11.3, the oblivion behaviour must be systematized by its static and dynamic components. According to the recent values of $g'((wi, wj), t)$ the following four types of edges can be distinguished:

1. **Permanent edges, with** $g'((w_i, w_j), t) > \alpha$,
   typically $\alpha \geq 0.85$ using the Dice coefficient for the edge weights in the co-occurrence graph, are those edges that model permanent knowledge that is no longer subject to forgetting processes in which in each case $g'((w_i, w_j), t) := g((w_i, w_j), t)$ is.
2. **Hidden edges, with** $g'((w_i, w_j), t) < \epsilon$,
   typically $\epsilon \leq 0.05$ were actually forgotten and should not longer be used for calculations in the co-occurrence graph. Nodes that are only adjacent with hidden edges also get the hidden status. On these edges, further reductions of $g'((w_i, w_j), t)$ are applied to these edges, but most importantly a memory process made possible to bring these edges back from 'oblivion'.
3. **Regular edges, with** $\epsilon < g'((w_i, w_j), t) < \alpha$,
   i.e. edges apply to the forgetting and remembering processes in full.
4. **Existing edges, with** $g'((w_i, w_j), t) > \epsilon$,
   i.e. all the edges that are used in a co-occurrence graph at a given time to be considered for further calculations and features.

After processing a document at time t, updates to the state of the co-occurrences graph are possible and needed. The following three changes are possible due to the described dynamics and essential:

1. **Oblivious or forgotten edges,**
   are those edges with $(g'((w_i, w_j), t) > \epsilon) \wedge (g'((w_i, w_j), (t+1) \leq \epsilon)$, i.e. which are transformed into hidden edges and taken out of the calculation of further features.
2. **Recalled or restored edges,**
   are those edges with $(g'((w_i, w_j), t) < \epsilon) \wedge (g'((w_i, w_j), (t+1) > \epsilon)$ that are transferred from the hidden state to the status of existing edges (normal or permanent). It makes sense to consider that the applied Δ should be chosen such that every hidden edge is at least transformed back into a regular edge when the corresponding co-occurrence re-appears.
3. **Updated edges,**
   are those edges with $(\epsilon < g'((w_i, w_j), t) < \alpha) \wedge (g'((w_i, w_j), (t+1) > g'((w_i, w_j), (t+1)))$, i.e. all normal edges whose edge weight increases due to the re-reading of a co-occurrence in the current document. Normal edges can become permanent, if due to the update $g'((w_i, w_j), (t+1)) > \alpha$.

With these definitions and set calculations now first experiments with oblivious co-occurrences and co-occurrence graphs shall be made and described in the following section.

11.3 Experimental Results

11.3.1 Setup

First experiments have been made using a news corpus of 1000 articles from the newspaper *Die Welt*. For calculation, a notebook with 2.3 GHz and an intel i5 dual-core processor with $8GB$ RAM should be used. Oblivion rates λ were chosen by 0.01, 0.001 and 0.0001. First experiences made exhibit huge simulation times in the area of 3 weeks to compute the given, 1000 documents and calculate for each time step the respective oblivion values as well as the updated graph. Since a co-occurrence graph is a huge structure, i.e. may have up to 500,000 nodes and a number of edges raising proportionally to the number of nodes with the power of 2, acceleration for the computation must be found. Therefore an update in the implementation was made as followed:

- All edges are storing additional values besides g and g'. Mostly the time of the last update was added. recent values for g' are only calculated if the corresponding occurrence appear again or the edge was needed for any other purpose.
- For evaluation purposes, the updated co-occurrence graph was only counted after a fixed, configurable set of time steps.

With these modifications, reasonable computation times, independent to λ and approximately linear growing with the number of computer documents as shown in Fig. 11.4 were obtained.

11.3.2 Results and Discussion

Since the oblivion process is connected in our model with the weight of the edges, mostly the number of the different groups of edges is important for the evaluation. Nodes once added may get isolated, if all adjacent edges are forgotten, however still exist as known and understood elements (what perfectly corresponds to the human brain model).

Figure 11.5 shows the number of existing edges, depending on different oblivion values and decomposed into the different groups of edges. It is to be seen that for a wide range of oblivion values of λ (from 0.01..0.0001) the curves remain in a stable, converging range. Only in case of strong oblivion in an interval between $\lambda = 0.01.0.001$ much more values are forgotten than really new learnt and the co-occurrence graph (the represented individual) does not represent new knowledge.

Figure 11.6 show the part of each of the four edge group among the entire set of edges characterising different good learning systems with different abilities to neglect or process fakes or outdated knowledge. Probably parameters of λ between 0.001 (b) and 0.001 (c) represent the reality is the best manner, while to high oblivion rates in (a) result in very small co-occurrence graphs, which mostly contain a frequently

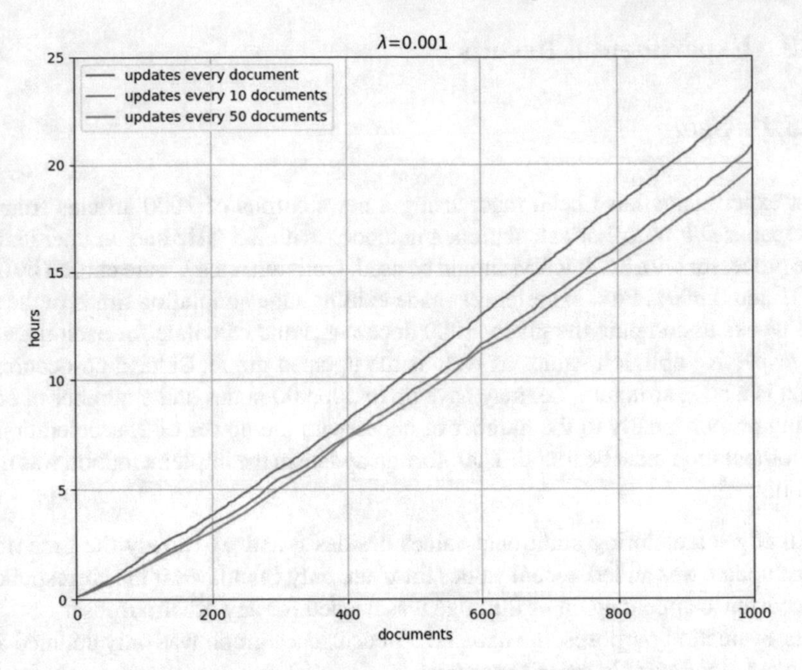

Fig. 11.4 Calculation times for the document processing with oblivion

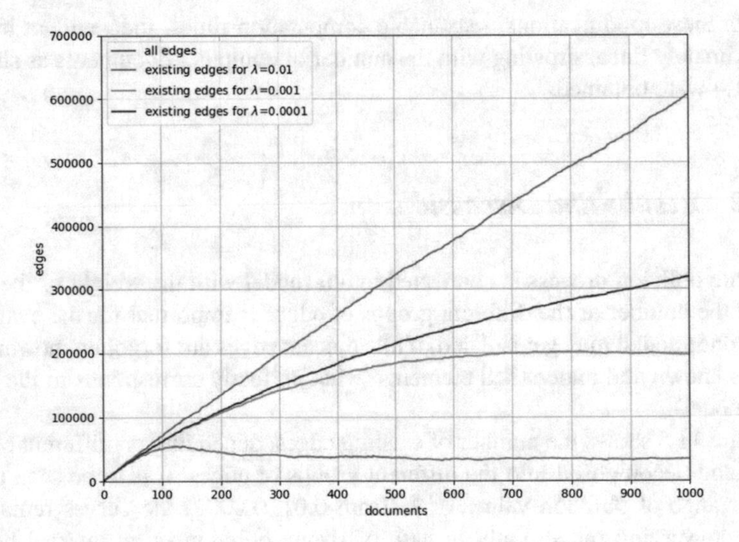

Fig. 11.5 Existing edges in the co-occurrence graph depending on λ

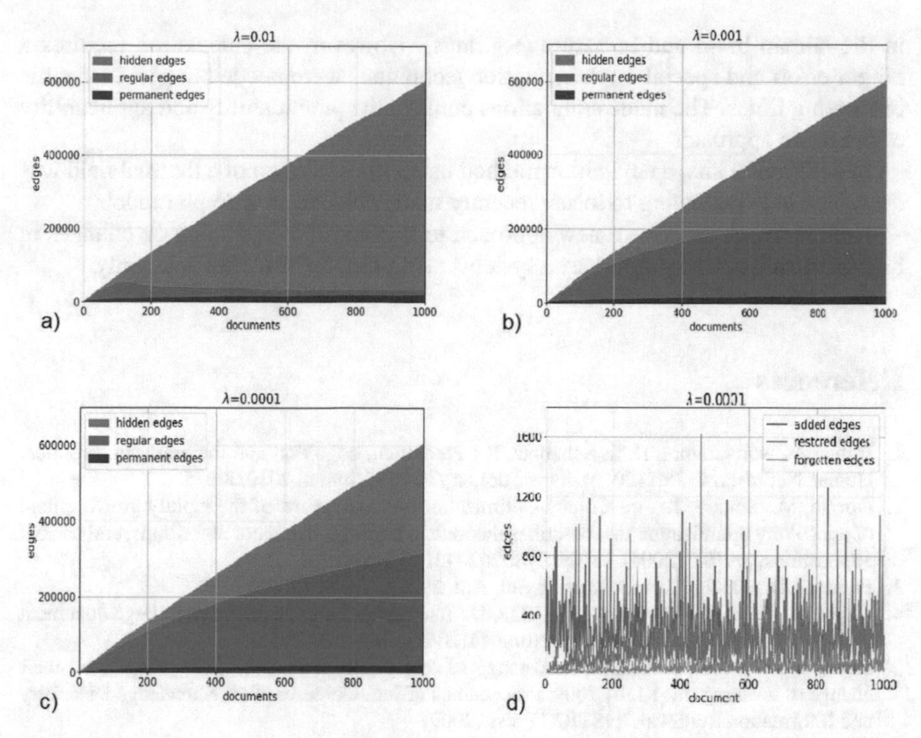

Fig. 11.6 Statistics on all edge groups for **a** $\lambda = 0.01$, **b** $\lambda = 0.001$ **c** $\lambda = 0.0001$ and **d** balance between added, forgotten and recovered edges

used, minimal amount of words and their co-occurrences (what might be meaningful, if frequent activities or opinions etc. of an individual shall be selected). (b) represents a learner with limited abilities (i.e. who's words and connection are limited) while (c) with the lowest oblivion rate represents a permanent learner with oblivion of old, unwanted elements, only.

Finally, Fig. 11.6d shows that—with $\lambda = 0.0001$ best to be seen– the also good balance between new added or recovered and forgotten edges will appear. Only at the beginning of the reading (learning) process, the growth process is predominating, of course. In addition, the random character of the process is well to be seen in Fig. 11.6d.

11.4 Conclusion

A method for digital oblivion has been presented, which can contribute to model continuous oblivion and update progress for co-occurrence graphs to actual events and changes in language. The proposed calculations are similar to the processes in the human brain and in nature (e.g. ants). However, the calculation requires a bigger effort and special implementation techniques were needed to get reasonable processing times. The made simulations confirm the practicability and applicability of the made approach.

In addition to this, a navigation method using the analogon of a thermal field was described in [8] allowing to locate recently made changes in a graph model.

Both methods describe a new approach to handle time depending on changes in big decentralised and graph-based systems using locally available data, only.

References

1. Bubic, A., Von Cramon D.Y., Schubotz, R.: Prediction, cognition and the brain. In: Frontiers Human Neurosci. **4**, 25ff (2010). https://doi.org/10.3389/fnhum.2010.00025
2. Dorigo, M., Stützle, T.: Ant Colony Optimization—An overview of the rapidly growing field of ant colony optimization that describes theoretical findings, the major algorithms, and current applications, pp. 7ff, (2004). ISBN: 9780262042192 319
3. Fields, R.D.: Making memories stick. Sci. Am. **292**(2), 75–81 (2005)
4. Heyer, G., Keim, D., Teresniak, S., Oelke, D.: Interactive explorative search in large document collections published. Database Spectrum **11**(2011), 195–206 (2011)
5. Heyer, G., Holz, F., Teresniak, S.: Change of topics over time and tracking topics by their change of meaning. In: KDIR 2009: Proceedings of Intl. Conference on Knowledge Discovery and Information Retrieval, INSTICC Press (2009)
6. Hloch, M., Kubek, M., Unger, H.: A survey on innovative graph-based clustering algorithms. In: The Autonomous Web (2021)
7. MacQueen, J.B.: Some methods for classification and analysis of multivariate observations. In: Proceedings of the 5th Berkeley Symposium on Mathematical Statistics and Probability, vol. 1, p. 281ff. University of California Press (1967)
8. Wulff, M.: Untersuchungen zur Strukturbildung in P2P-Netzen durch "Random Walker". Dissertation, Universität Rostock, Dept. of Computer Science (2006)

Printed in the United States
by Baker & Taylor Publisher Services